Biology of Nematodes: Current Studies

Papers by
R. Behme, M. A. Fernando, H. Hu Chow,
W. P. Rogers, Morton Rothstein,
Frank Stringfellow, Kazuo Yasuraoka,
J. Pasternak, W. F. Hieb, Anthony
J. Nappi, George J. Jackson, Irene
L. Riding, T. P. Bonner, George O.
Poinar, Jr., W. T. Springer, F. Donald
Tibbitts, Francis G. Tromba, Aaron
Goldberg, Andrew Covalcine, P. A. J.
Ball, D. R. Stoltz, Karen R. Clarke,
B. A. Obiamiwe, Raymond Cypess, J. S.
Pearse, R. J. Hudson, J. P. Carter, et al.

MSS Information Corporation
655 Madison Avenue, New York, N.Y. 10021

TABLE OF CONTENTS

CREDITS AND ACKNOWLEDGEMENTS

Ball, P. A. J.; and Ann Bartlett, "Serological Reactions to Infection with *Necator americanus,*" *Transactions of the Royal Society of Tropical Medicine and Hygiene,* 1969, 63:362-369.

Ball, P. A. J.; and Ann Bartlett, "A Method of Labelling Parasitic Nematodes with Carbon 14," *Transactions of the Royal Society of Tropical Medicine and Hygiene,"* 1969, 63:19.

Behme, R.; and J. Pasternak, "DNA Base Composition of Some Free-living Nematode Species," *Canadian Journal of Genetics and Cytology,* 1969, 11:993-1000.

Bonner, T. P.; F. J. Etges; and M. G. Menefee, "Changes in the Ultrastructure of *Nematospiroides dubius* (Nematoda) Intestinal Cells during Development from Fourth Stage to Adult," *Zeitschrift Zellforschung und Mikroskopische Anatomie,* 1971, 119:526-533.

Bonner, T. P.; and Paul P. Weinstein, "Ultrastructure of the Hypodermis during Cuticle Formation in the Third Molt of the Nematode *Nippostrongylus brasiliensis,*" *Zeitschrift Zellforschung und Mikroskopische Anatomie,* 1972, 126:17-24.

Carter, J. P.; R. Vanderzwaag; W. J. Darby; E. J. Lease; F. H. Lauter; B. W. Dudley; E. G. High; D. J. Wright; and T. Murphree, "Nutrition and Parasitism Among Rural Pre-school Children in South Carolina," *Journal of the National Medical Association,* 1970, 62:181-191.

Chow, H. Hu; and J. Pasternak, "Protein Changes during Maturation of the Free-living Nematode, *Panagrellus silusiae,*" *Journal of Experimental Zoology,* 1969, 170:77-84.

Clarke, Karen, R., "The Effect of a Low Protein Diet and a Glucose and Filter Paper Diet on the Course of Infection of *Nippostrongylus brasiliensis,*" *Parasitology,* 1968, 58:325-339.

Covalcine, Andrew; Sei Yoshimura; and Russell F. Krueger, "Application of the Bacteriological Pour-plate to Facilitate Mouse Pinworm Counts," *The Journal of Parasitology,* 1971, 57:854.

Cypess, Raymond, "Artificial Production of Acquired Immunity in Mice by Footpad Injections of a Crude Larval Extract of *Nematospiroides dubius,*" *The Journal of Parasitology,* 1970, 56:320.

Cypess, Raymond, "Demonstration of Immunity to *Nematospiroides dubius* in Recipient Mice Given Spleen Cells," *The Journal of Parasitology,* 1970, 56:199-200.

Fernando, M. A., "Hemoglobins of Parasitic Nematodes. II. Electrophoretic Analysis of the Multiple Hemoglobins of Adults and Developmental Stages of the Rabbit Stomach Worm, *Obeliscoides cuniculi,*" *The Journal of Parasitology,* 1969, 55:493-497.

Goldberg, Aaron, "Development and Survival on Pasture of Gastrointestinal Nematode Parasites of Cattle," *The Journal of Parasitology,* 1968, 54:856-862.

Hieb, W. F.; and Morton Rothstein, "Isolation from Liver of a Heat-stable Requirement for Reproduction of a Free-living Nematode," *Archives of Biochemistry and Biophysics,* 1970, 136:576-578.

Hieb, W. F.; and E. L. R. Stokstad, "Heme Requirement for Reproduction of a Free-living Nematode," *Science,* 1970, 168:143-144.

Hudson, R. J.; P. J. Bandy; and W. D. Kitts, "*In Vitro* Detection of Homocytotropic Antibody in Lungworm-infected Rocky Mountain Bighorn Sheep," *Clinical and Experimental Immunology,* 1971, 8:345-354.

Jackson, George J.; and Phyllis C. Bradbury, "Cuticular Fine Structure and Molting of *Neoplectana glaseri* (Nematoda), after Prolonged Contact with Rat Peritoneal Exudate," *The Journal of Parasitology,* 1970, 56:108-115.

Nappi, Anthony J.; and John G. Stoffolano, Jr., "*Heterotylenchus autumnalis:* Hemocytic Reactions and Capsule Formation in the Host, *Musca domestica,*" *Experimental Parasitology,* 1971, 29:116-125.

Obiamiwe, B. A., "The Life Cycle of *Romanomermis* sp. (Nematoda: Mermithidae) A Parasite of Mosquitoes," *Transactions of the Royal Society of Tropical Medicine and Hygiene,* 1969, 63:18-19.

Pasternak, J.; and M. R. Samoiloff, "The Effect of Growth Inhibitors on Postembryonic Development in the Free-living Nematode, *Pangrellus silusiae,*" *Comparative Biochemistry and Physiology,* 1970, 33:27-38.

Pearse, J. S.; and R. W. Timm, "Juvenile Nematodes (*Echinocephalus pseudouncinatus*) in the Gonads of Sea Urchins (*Centrostephanus coronatus*) and Their Effect on Host Gametogenesis," *Biological Bulletin*, 1971, 140:95-103.

Poinar, George O., Jr.; and Ruth Leutenegger, "Ultrastructural Investigation of the Melanization Process in *Culex pipiens* (Culicidae) in Response to a Nematode," *Journal of Ultrastructure Research,* 1971, 36:149-158.

Riding, Irene L., "Microvilli on the Outside of a Nematode," *Nature*, 1970, 226:179-180.

Rogers, W. P., "The Function of Leucine Aminopeptidase in Exsheathing Fluid," *The Journal of Parasitology*, 1970, 56:138-143.

Rothstein, Morton, "Nematode Biochemistry. X. Excretion of Glycerol by Free-living Nematodes," *Comparative Biochemistry and Physiology*, 1969, 30:641-648.

Springer, W. T.; Joyce Johnson; and W. M. Reid, "Transmission of Histomoniasis with Male *Heterakis gallinarum* (Nematoda)," *Parasitology*, 1969, 59:401-405.

Stoltz, D. R.; and I. K. Barker, "Bladder Tumorigenesis," *Science*, 1970, 168:1121-1122.

Stringfellow, Frank, "Functional Morphology and Histochemistry of Structural Proteins of the Genital Cone of *Cooperia punctata* (Von Linstow, 1907) Ransom, 1907, A Nematode Parasite of Ruminants," *The Journal of Parasitology*, 1969, 55:1191-1200.

Tibbitts, F. Donald; and Bert B. Babero, "*Ascaridia galli* (Schrank, 1788) from the Chukar Partridge, *Alectoris chukar* (Gray), in Nevada," *The Journal of Parasitology*, 1969, 55:1252.

Tromba, Francis G.; and Frank W. Douvres, "Survival of Juvenile and Adult *Stephanurus dentatus in Vitro*," *The Journal of Parasitology*, 1969, 55:1050-1054.

Yasuraoka, Kazuo; and Paul P. Weinstein, "Effects of Temperature on the Development of Eggs of *Nematospiroides dubius* under Axenic Conditions Relative to *in Vitro* Cultivation," *The Journal of Parasitology*, 1969, 55:44-50.

PREFACE

Nematodes are a large, complex group whose phylogenetic relationships with other major metazoan phyla are far from clear. Much of the research on nematodes is related to their frequent role as more or less pathological parasites of man and other mammals.

Including only those papers published within the last 2-3 years, this volume deals with the molecular and cellular biology of nematodes, their physiology, and ultrastructure. Host-nematode interactions, their immune response, and the occurence and detection of nematodes are among the topics discussed.

Molecular and Cell Biology of Nematodes

DNA BASE COMPOSITION OF SOME FREE-LIVING NEMATODE SPECIES

R. Behme and J. Pasternak

Introduction

The base composition of the DNA molecules of an extensive assortment of organisms have been determined within the past 10 years (Sueoka, 1964; Shapiro, 1969). However, there are conspicuous gaps in the available information of the mole per cent of guanine plus cytosine (% GC) of the DNA of some important biological groups. For example, the protostomes, especially the phylum Aschelminthes, class Nematoda, have been rarely examined from this point of view. The estimated number of species of nematodes is considerable (Hyman, 1951). Yet, as far as we are aware, the base composition of the DNA of only one nematode, *Ascaris lumbricoides*, Linnaeus, 1758, is known (Kaulenas and Fairbairn, 1968; Bielka, Schultz and Böttger, 1968).

To partially remedy this situation, we have determined the % GC of DNA isolated from five free-living nematodes by CsCl density gradient equilibrium centrifugation and thermal transition temperature (Tm) studies. Studies of the properties of DNA molecules in related organisms have provided a means for assessing taxonomic affinities (Marmur, Falkow and Mandel, 1963; Storck, 1966; McCarthy, 1965). In the present investigation the role of DNA base analysis information as an indicator of genetic relatedness was demonstrated. Two nematode species in the same genus were found to have identical DNA base compositions. Breeding experiments revealed that these forms are cross-fertile and the offspring are fully fertile.

Materials and Methods

Culturing of Organisms

The various nematode species were kindly supplied by Dr. A. C. Coomans (*Panagrellus silusiae*), Dr. H. Gysels (*Panagrellus silusiae*), Dr. E. Yarwood

10

(*Panagrellus redivivus, Rhabditis anomala, Caenorhabditis briggsae, Turbatrix aceti*). Dr. M. Rothstein (*Panagrellus redivivus, Caenorhabditis briggsae, Turbatrix aceti*) and Dr. B. Zuckerman (*Panagrellus redivivus*). The nematodes were grown on 3.4% (W/v) Czapek Dox agar (Oxoid) as described previously (Chow and Pasternak, 1969; Samoiloff and Pasternak, 1969). The *T. aceti* cultures were supplemented with 0.5% glacial acetic acid. *P. redivivus, P. silusiae,* and *C. briggsae* were kept at 26°C. *R. anomala* and *T. aceti* were maintained at 20°C.

Isolation and Analysis of DNA

The nematodes were freed of agar pieces and debris by straining through four layers of cotton gauze, harvested by low speed centrifugation (1400 $\times$ g), and washed at least five times with 1 $\times$ SSC (SSC is 0.15M NaCl, 0.015M trisodium citrate (pH 7.0); 2 $\times$ SSC is twice the concentration of SSC, etc.). The final washed pellet contained about 0.3 to 0.5 ml of packed nematodes. To the final volume of packed nematodes 2 $\times$ SSC was added to make a 3-ml suspension.

The nematodes were homogenized by vigorous agitation with an equal volume of glass beads (diameter about 0.5 mm) in a Mickle disintegrator for 15 min at 4°C. The procedure for extracting DNA was similar to that described by Marmur (1961) except that 2 $\times$ SSC replaced saline-EDTA and a deproteinization step with self-digested pronase (50 μg/ml) for 2 hr at 37°C followed the RNase treatment. In some instances, the DNA solution was dialyzed against polyethylene glycol for several hours at room temperature prior to the pronase and RNase treatments. A typical ultraviolet spectrum was observed with the nematodal DNA preparations with the absorbance ratios ($_{260}$/A$_{280}$ and A$_{260}$/A$_{230}$) usually equal to 2.0.

The buoyant density of each isolated DNA was determined by CsCl isopycnic density gradient centrifugation with a Spinco (Model E) analytical ultracentrifuge operated at 44770 RPM at 20°C for about 20 hr. DNA of *Aerobacter aerogenes* ($\rho = 1.715$ g/cm^3; Behme, 1969) was added as the density standard. Ultraviolet and schlieren photographs were taken after 20 hr of centrifugation and scanned with a Gilford Linear Transport unit with a silt opening of 0.5 mm. Buoyant density of a DNA band was used to calculate the % GC content by the formula of Schildkraut, Marmur and Doty (1962).

Thermal denaturation of the DNA in 0.25 $\times$ SSC or 1 $\times$ SSC conformed to the procedure described by Marmur and Doty (1962). The temperature increments were controlled by a Neslab linear temperature programmer. The temperature in a cuvette containing solvent was monitored by a Tri-R electronic thermometer. All optical density measurements were made at 260 mμ with a Gilford photometer and Beckman monochromator. A linear equation was derived using DNA from organisms (*Paramecium aurelia, Aerobacter aerogenes* and *Saccharomyces cerevisae*) with known base composition that related the thermal denaturation temperature (Tm) of DNA in 0.25 $\times$ SSC with the % GC content. The absorbance readings were corrected for dilution changes due to increasing temperatures (Mandel and Marmur, 1968).

11

Density Gradient Ultracentrifugation

When the DNA was analyzed by analytical ultracentrifugation in CsCl density gradients the banding profiles shown in Fig. 1 were observed. The DNA band with a buoyant density of 1.715 g/cm³ is the reference DNA extracted from the bacteria *Aerobacter aerogenes*. The buoyant densities of the DNA of the nematodes examined range from 1.695 g/cm³ (36% GC) for *Caenorhabditis briggsae* to 1.703 g/cm³ (44% GC) for both *Panagrellus silusiae* and *P. redivivus*.

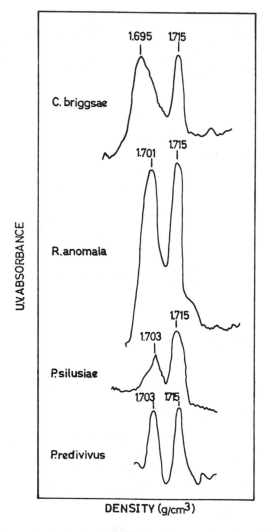

Fig. 1. Densitometric tracings of ultravoilet photographs of analytical CsCl density gradient of the DNA from four species of free-living nematodes. *Aerobacter aerogenes* DNA, $\rho = 1.715$ g/cm³, was used as a marker.

No evident 'satellite' components were found in four of the five species examined. The one exception is the DNA preparation from *Turbatrix aceti* Two bands, in addition to the marker DNA, with densities 1.688 g/cm³ and 1.699 g/cm³ were observed in UV photographs (Fig. 2).

When viewed with the schlieren optical system the material banding at 1.688 g/cm³ produced a large disturbance in the refractive index gradient, whereas the material at 1.699 g/cm³ did not (Fig. 3A). The schlieren refractive index pattern observed in Fig. 3A indicates the presence of polysaccharide material (Brunk and Hanawalt, 1966; Edelman, *et al.*, 1967). To confirm this contention aliquots of the DNA preparation from *T. aceti* were treated with α-amylase (250 μg/ml) or DNase (100 μg/ml) each at 37°C for 1 hr. Prior to the DNase treatment the preparation was dialyzed against 0.15M NaCl. The band at 1.688 g/cm³ was eliminated by α-amylase treatment (Figs. 2 and 3B) and was resistant to DNase degradation. On the other hand, the band at 1.699 g/cm³ was insensitive to α-amylase treatment and removed by DNase (Fig. 2). Thus, the DNA composition of *T. aceti* is 40% GC and the additional band is polysaccharide. The mean base composition (% GC) of each nematode species derived from the buoyant density data is shown in Table 1.

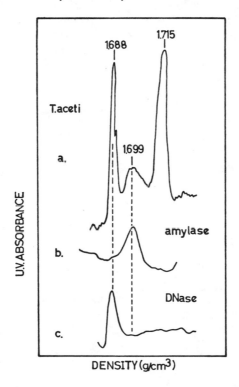

Fig. 2. Densitometric tracings of ultraviolet photographs of analytical CsCl density gradient of *T. aceti* DNA. (a) Complete DNA preparation with marker DNA added. (b) The same nematode DNA preparation as in (a) after treatment with α-amylase. (c) The same nematode DNA preparation as in (a) after treatment with deoxyribonuclease. The DNA from *Aerobacter aerogenes* (ρ = 1.715 g/cm³) was used as the reference in (a) but was not used in either (b) or (c).

13

Thermal Denaturation

Thermal transition values (Tm) were determined for four of the five nematode species. It was not possible to obtain reliable thermal denaturation data with DNA preparations from *T. aceti*. Figure 4 gives examples of the thermal transition curves for three species. The melting profiles showed a regular S-shape indicating a uniform distribution of the DNA molecules. The mean base composition was calculated from the Tm by the formula % GC = (Tm-69.3) 2.44 (Marmur and Doty, 1962) when 1 × SSC was used as the solvent. When the nematode DNA was dissolved in 0.25 × SSC the formula % GC = (1.65 Tm) −89.2 was used. The DNA from *P. silusiae* and *P. redivivus* were heated in 0.25 × SSC, and a Tm of 80.5° was recorded. The DNA from *R. anomala* (Tm = 86°C) and *C. briggsae* (Tm = 85°C) were heated in 1 × SSC. The Tm of DNA changes in a linear manner with the logarithm of the salt concentration. The more dilute the solvent the lower

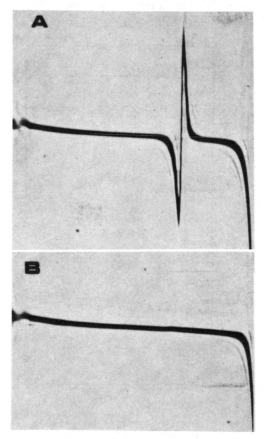

Fig. 3. Analytical CsCl density gradient of *T. aceti* DNA photographed with the schlieren optical system (A) *T. aceti* DNA preparation. The schlieren peak coincides with density 1.688 g/cm³. (B) *T. aceti* DNA preparation treated with α-amylase. The top of the gradient is to the right.

the Tm value. The mean base composition of the nematode DNA calculated from the Tm values are presented in Table 1.

Mating of P. silusiae *and* P. redivivus

The CsCl ultracentrifugation and thermal denaturation studies showed that the DNA from *P. silusiae* and *P. redivivus* have the same % GC content. To test whether or not this was fortuitous, reciprocal single pair matings of virgin animals were undertaken. In 81 out of 82 matings between *P. silusiae* and *P. redivivus* a fertile FI generation was obtained. To ensure that the original stocks were not contaminated, new cultures of both species were acquired and the matings were repeated. All of the ensuing interspecies crosses (63/63) gave fertile offspring. These results indicate that the forms designated *P. redivivus* and *P. silusiae* are genetically identical and, therefore, they should not be considered as two separate species.

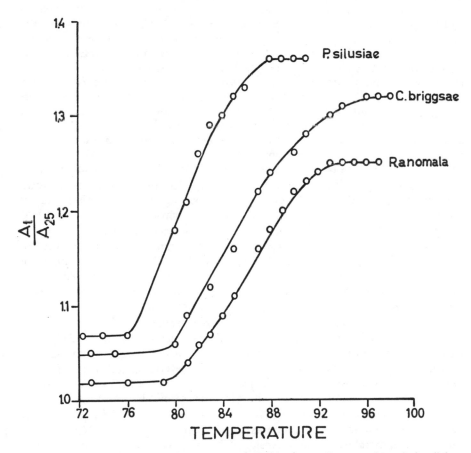

Fig. 4. Thermal denaturation curves of DNA from three species of free-living nematodes. Relative absorbance, corrected for thermal expansion, measured at increasing temperatures. DNA from *P. silusiae* was heated in 0. 25 × SSC; the DNA from both *R. anomala* and *C. briggsae* were heated in 1 × SSC.

15

TABLE I
Mean base composition (% GC) of nematode DNA

Organism	From buoyant density	From thermal denaturation
Panagrellus silusiae	44	44
Panagrellus redivivus	44	44
Caenorhabditis briggsae	36	39
Rhabditis anomala	42	42
Turbatrix aceti	40	–

Discussion

The mean base composition of the DNA of the five species of nematodes examined in the present study covered a narrow range (36-44% GC). Kaulenas and Fairbairn (1968) reported that the DNA from *Ascaris lumbricoides*, a parasitic nematode, has a % GC content of 39.4. According to Bielka, Schultz and Böttger (1968) the base composition of nuclear DNA of eggs and gastrulae of *Ascaris* is 37.7%. Base composition information alone is not sufficient evidence upon which taxonomic affinities can be based or substantiated. The % GC values, however, do provide a first approximation toward the genetic relationships among the nematodes. Although identity of base composition does not necessarily indicate the existence of identical nucleotide sequences of significant length, extensive sequence homology does require similarity of mean base composition. In this context, it was instructive to find that *P. silusiae* and *P. redivivus* were cross-fertile after the physical methods had shown them to have the same DNA base composition. Similar analyses may be helpful in the study of the systematics of other closely related free-living nematodes.

The close agreement of the values of % GC determined from the buoyant density in CsCl and thermal denaturation experiments indicate that there is no significant replacement of the normal bases in nematode DNA.

The presence of the polysaccharide contamination in DNA extracts has been documented by Edelman *et al.*, (1967) and Brunk and Hanawalt (1966). *T. aceti* was the only organism in the present study to show a minor component that could be construed as a satellite band. This component ($\rho = 1.688$ g/cm³) gave a distinctive light scattering pattern when observed with the schlieren optical system and was eliminated by α-amylase treatment.

These observations demonstrate that the material was a polysaccharide. No attempt was made to characterize this band further. Commercially prepared glycogen centrifuged in CsCl bands at 1.675 g/cm³ (Mandel and Marmur, 1968). On the other hand, the polysaccharide material found in the *T. aceti* DNA fraction bands at 1.688 g/cm³. Although the *T. aceti* material is sensitive to α-amylase degradation identity with glycogen cannot be presupposed.

Beilka, Schultz and Böttger (1968) found that *ascarid* DNA preparations have two UV absorbing components in addition to the main DNA fraction. One of these 'extra' bands has a density of 1.685 g/cm³. This density is similar to the polysaccharide band found in the *T. aceti* extract. It is possible that the component banding at 1.685 g/cm³ from the *ascarid* DNA fraction is a polysaccharide. Studies on DNA extracted from purified mitochondria may reveal the source of one of the satellite bands found in *Ascaris*. In the present study there was no evidence of a DNA fraction differing from the principal DNA. This observation suggests that mitochondrial DNA and nuclear DNA may have similar base compositions. However, reliable resolution of this particular problem requires that mitochondrial DNA be prepared from a mitochondrial fraction.

Acknowledgements

We thank the Department of Microbiology, University of Guelph, Guelph, Ontario, for the use of the Model E. centrifuge facility and Dr. J. Douglas for his excellent assistance. Mrs. L. Anderson proficiently assisted in the extraction of some of the DNA. This study was supported by a National Research Council of Canada grant (A-3491).

References

Behme, R. J. 1969. Deoxyribonucleic acid relationships among symbionts of *Paramecium aurelia*. Ph.D. Thesis. Indiana University, Bloomington, Indiana.

Bielka, H., Schultz, I., and Böttger, M. 1968. Isolation and properties of DNA from eggs and gastrulae of *Ascaris lumbricoides*. Biochim. Biophys. Acta **157**: 209-212.

Brunk, C. F., and Hanawalt, P. C. 1966. Glycogen satellite bands in isopycnic CcCl gradients. Exp. Cell Res. **42**: 406-408.

Chow, H. Hu, and Pasternak, J. 1969. Protein changes during maturation of the free-living nematode, *Panagrellus silusiae*. J. Exp. Zool. **170**: 77-84.

Edelman, M., Swinton, D., Schiff, J. A., Epstein, H. T., and Zelden, B. 1967. Deoxyribonucleic acid of the blue-green algae (Cyanophyta). Bacteriol. Rev. **31**: 315-331.

Hyman, L. 1951. The pseudocoelomate bilateria, *In* The invertebrates: Acanthocephala, Aschelminthes, and Entoprocta. Vol. III. McGraw-Hill Book Company, New York. 572 pp.

Kaulenas, M. S., and Fairbairn, D. 1968. RNA metabolism of fertilized *Ascaris lumbricoides* eggs during uterine development. Exp. Cell Res. **52**: 233-251.

Mandel, M., and Marmur, J. 1968. Use of ultraviolet absorbance-temperature profile for determining the guanine plus cytosine content of DNA, pp. 195-206. *In* L. Grossman and K. Moldave [ed.] Methods in enzymology, nucleic acids. Vol. XII, part B. Academic Press Inc., New York.

Marmur, J. 1961. A procedure for the isolation of deoxyribonucleic acids from microorganisms. J. Mol. Biol. **3**: 208-218.

Marmur, J., and Doty, P. 1962. Determination of the base composition of DNA from its thermal denaturation temperature. J. Mol. Biol. **5**: 109-118.

Marmur, J., Falkow, S., and Mandel, M. 1963. New approaches to bacterial taxonomy. Ann. Rev. Microbiol. **17**: 329-372.

McCarthy, B. J. 1965. The evolution of base sequences in polynucleotides. Progr. Nucleic Acid Res. Mol. Biol. **4**: 129-160.

Samoiloff, M. R., and Pasternak, J. 1969. Nematode morphogenesis: fine structure of the molting cycles in *Panagrellus silusiae* (de Man 1913) Goodey 1945. Can. J. Zool. **47**: 639-644.

Schildkraut, C. L., Marmur, J., and Doty, P. 1962. Determination of the base composition of deoxyribonucleic acid from its buoyant density in CsCl. J. Mol. Biol. **4**: 430-443.

Shapiro, H. S. 1968. Distribution of purines and pyrimidines in deoxyribonucleic acids, Section H, pp. H30-H51. *In* H. A. Sober [ed.] Handbook of biochemistry. Selected data for molecular biology. Chemical Rubber Company.

Storck, R. 1966. Nucleotide composition of nucleic acids in fungi. J. Bacteriol. **91**: 227-230.

Sueoka, N. 1964. Compositional variation and heterogeneity of nucleic acids and protein in bacteria, pp. 419-443. *In* I. C. Gunsalus and R. Y. Stainer [ed.] The bacteria, Vol. 5. Academic Press Inc., New York.

HEMOGLOBINS OF PARASITIC NEMATODES. II. ELECTROPHORETIC ANALYSIS OF THE MULTIPLE HEMOGLOBINS OF ADULTS AND DEVELOPMENTAL STAGES OF THE RABBIT STOMACH WORM, OBELISCOIDES CUNICULI*

M. A. Fernando

In a recent paper Fernando (1968) described the occurrence and distribution of hemoglobin and other hematin compounds within the tissue of *Obeliscoides cuniculi.* Hemoglobin was detected in the perienteric fluid, in the tissues of the body walls, and in the intestinal cells of adult worms. Preliminary experiments indicated the presence of several hemoglobin components distributed in the above tissues.

This paper presents an electrophoretic analysis of these multiple hemoglobins, their tissue specificity, and their developmental sequence.

MATERIALS AND METHODS

The adults and the appropriate larval stages of *O. cuniculi* and the necessary tissue components of adult females including their perienteric fluid were obtained as described by Fernando (1968).

Whole worms or individual tissue components were homogenized in the appropriate buffer in an all-glass tissue grinder, the homogenate centrifuged at 1,500 g (maximum), the supernatant passed through an $0.8-\mu$ millipore filter, and the filtrate used in the electrophoretic analysis. Perienteric fluid was similarly treated and the filtrate used for the analysis of its hemoglobins. When necessary, these filtrates were concentrated either by ultra-filtration or according to Kohn's (1959) method. All operations were carried out at 4 C. The hemoglobin in the samples was usually converted into

* This work was supported in part by the National Research Council of Canada and by the Ontario Department of Agriculture and Food.

the carbon monoxide derivative for increased stability. However, oxyhemoglobin, methemoglobin, and methemoglobin cyanide, prepared according to Fernando (1968), were also studied.

Smithies (1955, 1959) horizontal and vertical starch gel electrophoretic techniques were used to analyze the worm hemoglobins. Hydrolyzed starch was obtained from Connaught Laboratories, Toronto, and used at a concentration of 12.6% unless otherwise stated. Electrophoresis was carried out at 350 to 400 volts, at 4 C, in Tris-EDTA Borate buffer giving a gel pH of 8.8. Samples were introduced on squares of Whatman 3 MM filter paper for electrophoresis in the horizontal direction and directly into the sample slots provided in the Büchler apparatus which was used for vertical starch gel electrophoresis. At the end of the run the gels were sliced, one half stained with amido black to detect protein and the other half stained with a peroxidase stain to detect the hemoglobin. Of the stains tested, O' dianisidine prepared according to Bodman (1960) proved to be the most sensitive and the reaction product was stable for several months. O' dianisidine was therefore used routinely to stain the hemoglobins separated on starch gels.

Unstained starch gels were made translucent and the various electrophoretic hemoglobin components analyzed spectrophotometrically as described by Gratzer and Beaven (1960). Spectrophotometric analysis was also performed on hemoglobin components eluted by freezing and thawing the appropriate portion of the starch gel.

Polyacrylamide disc electrophoresis was performed according to Davies (1964) using 7.5% gels and a buffer pH of 8.9. A stacking gel and a sample gel were used in the analysis and the migration front was marked with bromophenol blue. Electrophoresis was performed at 3 to 5 ma per tube giving adequate separation in 20 to 30 min at 4 C. The gels were stained either with

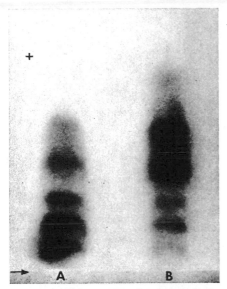

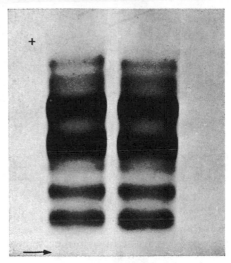

FIGURE 1. Pattern obtained on starch gel electrophoresis of (A) male and (B) female whole-worm homogenates of *Obeliscoides cuniculi.* Gel was stained with O' dianisidine to show the hemoglobins. Arrow represents origin.

FIGURE 2. Pattern obtained on disc electrophoretic analysis of female whole-worm homogenates. Gel stained with benzidine to show the hemoglobins. Arrow represents origin.

amido black or with a peroxidase stain. O' dianisidine stain which was used successfully on the starch gels gave poorer results on the acrylamide gels than the benzidine stain prepared as described by Nerenberg (1966).

Two-dimensional electrophoresis in varying concentrations of acrylamide gel was used to check for the presence of polymers among the different electrophoretic hemoglobin components. The method used was essentially that described by Raymond (1964) but the runs were made in the horizontal instead of in the vertical direction. Five per cent acrylamide was used for electrophoresis in the first direction and 8% acrylamide for electrophoresis in the second direction. The gels were stained either with amido black or with the benzidine stain.

RESULTS

Electrophoresis of the hemoglobins of adult *O. cuniculi*

The resolution on starch gel electrophoresis of the hemoglobins of homogenates of whole male and whole female *O. cuniculi* is shown in Figure 1. Six hemoglobin components are consistently resolved, all moving towards the anode at the pH of electrophoresis. The component nearest the origin has been designated

Hb-1 and that nearest the anode, Hb-6. Hb-1, -2, and -3 form the major components in male worms and Hb-4 and -5 form the major components in female worms. Electrophoresis of the hemoglobins of individual female worms showed that all six hemoglobin components were present in each individual. The concentration of hemoglobin in male worms was too low to permit such an analysis.

The heterogeneity of *O. cuniculi* hemoglobin and the difference between the male and the female hemoglobins persist whether the hemoglobin is converted to oxy-Hb, CO-Hb, methemoglobin, or methemoglobin cyanide. The difference noted in the electrophoretic mobility especially with change of their state of oxidation is in agreement with studies on human hemoglobins (Chernoff and Pettit, 1964). These alterations do not, however, change the fundamental heterogeneity of *O. cuniculi* hemoglobins. This suggests that the observed multiplicity of the hemoglobins is not the result of some of the hemoglobins being in a different oxidation state or combined with different ligands. This is confirmed by results obtained on spectrophotometric analysis of the different hemoglobin fractions on translucent starch gels.

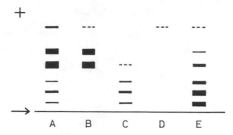

FIGURE 3. Diagrammatic representation of the distribution of the hemoglobin components in the tissues of adult female worms. A, Female whole worm homogenate; B, perienteric fluid; C, body wall homogenate; D, homogenate of intestine; E, male whole-worm homogenate.

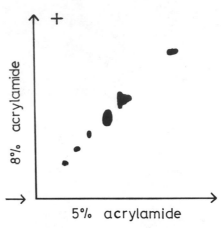

FIGURE 4. Diagrammatic representation of two-dimensional electrophoresis of the hemoglobins of female whole-worm homogenate on acrylamide gels.

The pattern obtained when hemoglobins of female worms are separated by polyacrylamide disc electrophoresis is shown in Figure 2. It is essentially similar to the separation obtained on starch gels except that two bands are present at the position of Hb-6.

Tissue-specific hemoglobin components

In order to ascertain whether the observed heterogeneity of *O. cuniculi* hemoglobin was due, at least in part, to tissue-specific components, the perienteric fluid, body walls, and intestines of adult females were separately electrophoresed.

The distribution of the hemoglobin components in the tissues of adult female worms is illustrated in Figure 3. Hb-1, -2, and -3 are present in the tissues of the body walls, Hb-4 and -5 and traces of -6 are present in the perienteric fluid, and Hb-6 is present, though not consistently, in intestinal homogenates. A trace of Hb-4 is occasionally detected in body wall homogenates that have not been adequately washed and is probably a contaminant. There are thus at least two tissue-specific groups of hemoglobins in adult *O. cuniculi*. These results also provide a possible explanation for the differences observed (Fig. 1) between males and females in the relative concentration of their hemoglobin components. Male worms probably contain a higher concentration of body wall hemoglobins as compared to their perienteric fluid hemoglobins and female worms a comparatively higher concentration of perienteric fluid hemoglobins.

Differences are noted in the oxyhemoglobin absorption spectrum between the two tissue-specific groups but within each group the spectrum remains the same from one electrophoretic component to the next. The peak in the Soret region was at 415 mμ for Hb-1, -2, and -3 and at 411 mμ for Hb-4, -5, and -6.

Molecular aggregates

Preliminary attempts were made to find out whether the heterogeneity within the two tissue-specific groups of hemoglobins could be the result of different degrees of stabilized molecular aggregation. The result of one such experiment—two-dimensional acrylamide gel electrophoresis—is represented diagrammatically in Figure 4. There is no differential retardation in electrophoretic mobility of any of the hemoglobin components by changes in the amount of acrylamide used to form the gels.

Ontogeny of *O. cuniculi* hemoglobins

Figure 5 is a diagrammatic representation of electrophoretic patterns of the hemoglobins of 4th-stage *O. cuniculi* larvae obtained approximately 6 days after host infection. The body wall hemoglobins seem to be the first to appear in male larvae and the perienteric fluid hemoglobins the first to appear in female

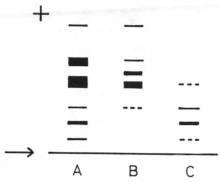

FIGURE 5. Diagrammatic representation of the electrophoretic patterns on starch gels of A, adult female hemoglobins; B, 4th-stage female larval hemoglobins; C, 4th-stage male larval hemoglobins.

larvae. An additional hemoglobin band is observed at this time in female larvae (Fig. 5) and persists until at least 12 days post-infection when the larvae are at the L5 stage. At this time all the hemoglobins found in adult worms could be detected in the L5 juveniles.

DISCUSSION

The results reported here confirm that the heterogeneity of *O. cuniculi* hemoglobin is not due to differences among individual worms in a given population. The differences in the oxyhemoglobin spectrum of the two tissue-specific groups clearly indicate that Hb-1, -2, and -3 differ from Hb-4, -5, and -6. The absence of differential retardation in electrophoretic mobility of these hemoglobin components with changes in the gel concentration indicate the absence of molecular aggregates within the two groups. Preliminary results obtained with sucrose gradient ultracentrifugation show a single hemoglobin peak representing the several electrophoretic fractions.

Multiple hemoglobins have been reported before from other parasitic nematodes. Hamada et al. (1962) found two hemoglobin components in ascaris body walls. These hemoglobins differed in their electrophoretic mobility, mobility in DEAE columns, and in their spectra. Villela and Ribeiro (1955) reported that the perienteric fluid of *Tetrameres confusa* contained three electrophoretically distinct hemoglobin components having different isoelectric points.

Preliminary analysis in our laboratory of the hemoglobins of other adult trichostrongyles indicates that they too contain several electrophoretically distinct hemoglobin components. *Ostertagia* sp. from cattle was found to contain five hemoglobin components electrophoretically identical with Hb-1, -2, -4, -5, and -6 of *O. cuniculi.*

Although the functional significance of these multiple forms of hemoglobin is not clear at present a study of their oxygen relationships, other biochemical properties such as isoelectric points, and the precise structural localization of the various hemoglobins may reveal their significance in the physiology of these nematodes. Such studies should also take into consideration the differences in the relative concentration of the various hemoglobins that exist, even from the earliest larval stages, between male and female *O. cuniculi.* Differences of this nature, although not reported so far, probably also exist in other parasitic nematodes.

ACKNOWLEDGMENTS

I wish to thank Dr. R. W. Stevens of the Department of Bacteriology, University of Guelph, for help with the initial experiments and the members of my own department for their help and advice given on many occasions. The skilled technical assistance of Mrs. S. Shelofsky is gratefully acknowledged.

LITERATURE CITED

BODMAN, J. 1960. Agar gel, starch block, starch gel and sponge rubber electrophoresis. *In* Ivor Smith (ed.), Chromatographic and Electrophoretic Techniques. Vol. II. Zone Electrophoresis. Interscience Publ. Inc., New York, p. 91–157.

CHERNOFF, A. I., AND N. M. PETTIT, JR. 1964. Some notes on the starch-gel electrophoresis of hemoglobins. J. Lab. Clin. Med. **63:** 290–296.

DAVIES, B. J. 1964. Disc electrophoresis. II. Methods and application to human serum proteins. Ann. N. Y. Acad. Sci. **121:** 404–427.

FERNANDO, M. A. 1968. Hemoglobins of parasitic nematodes. I. Spectra of the perienteric fluid hemoglobins of *Obeliscoides cuniculi* and the distribution of hematin compounds within its tissues. J. Parasit. **54:** 863–868.

GRATZER, W. B., AND G. H. BEAVEN. 1960. Transparent starch gels: Preparation, optical properties and application to hemoglobin characterisation. Clin. Chim. Acta **5:** 577–582.

HAMADA, K., T. OKAZAKI, R. SHUKUYA, AND K.

KAZIRO. 1962. Hemoglobins from *Ascaris lumbricoides*. I. Purification of hemoglobin from body-wall tissue and its spectral properties. J. Biochem. Tokyo **52**: 290–296.

KOHN, J. 1959. A simple method for the concentration of fluids containing protein. Nature **183**: 1055.

NERENBERG, S. T. 1966. Electrophoresis. F. A. Davis Company, Philadelphia, 272 p.

RAYMOND, S. 1964. Acrylamide gel electrophoresis. Ann. N. Y. Acad. Sci. **121**: 350–365.

SMITHIES, O. 1955. Zone electrophoresis in starch gels: Group variations in the serum proteins of normal human adults. Biochem. J. **61**: 629–641.

————. 1959. An improved procedure for starch gel electrophoresis: Further variations in the serum proteins of normal individuals. Biochem. J. **71**: 585–592.

VILLELA, G. G., AND L. P. RIBEIRO. 1955. Hemoglobins of the worm *"Tetrameres confusa."* Rev. Brasil. Biol. **15**: 383–390.

Protein Changes During Maturation of the Free-living Nematode, *Panagrellus silusiae*

H. HU CHOW AND J. PASTERNAK

During the postembryonic development of nematodes there are four moults. Customarily each moult marks the end of one stage and the beginning of another. However, it is not known whether each stage makes a unique contribution to the maturation of the worm. The general constitution of the body differs little from stage to stage except for increase in size. The only major feature that distinguishes the juvenile stages from the mature adult stage is the presence of complete reproductive organs in the latter. Moreover, the organism used in this study belongs to that set of nematodes which are purported to lack somatic cell division during maturation (Chitwood, '50). Accordingly, it was of interest to analyze whether certain proteins fluctuate from stage to stage during postembryonic development.

It is possible to initiate one round of synchronous growth from the first postpartum stage to young adulthood of the nematode *Panagrellus silusiae* (Samoiloff and Pasternak, '68). This procedure affords a convenient system for the study of protein changes during maturation. As will be shown, polyacrylamide disc electrophoresis of homogenates of specific stages reveal that each free-swimming stage in the life cycle has a discrete array of enzymatic proteins.

MATERIALS AND METHODS

Culture and isolation procedures

The nematode strain *Panagrellus silusiae* (de Man, 1913 Goodey, 1945) was obtained from Dr. A. C. Coomans (Rijksuniversiteit, Gent, Belgium). The worms were maintained xenically on 4.5% (w/v) modified Czapek Dox agar (Oxoid) medium in Petri plates at 27°C. *P. silusiae* is ovoviviparous and the first moult takes place *in utero*; the ensuing four stages are free-swimming. The first postembryonic stage, which is prepartum, is called the L1 stage and, in succession the next four stages are L2, L3, L4 and adult. The simplest criterion for defining a particular stage is the length of the worm (Gysels and van der Haegen, '62; Samoiloff and Pasternak, '68). The mean length of the four postpartum stages are 350 μ, 550 μ, 750 μ and 950 μ for L2, L3, L4 and young adult stages, respectively. The limits of each stage are not precisely known. Worms that fall within the ranges 300–450 μ, 451–650 μ, 651–850 μ and over

851 μ are designated to be in L2, L3, L4 and adult stages respectively.

Mass cultures from ten Petri plates were washed through seven layers of cheese-cloth to screen out most of the agar pieces. The extracted worms were harvested by centrifugation for two minutes at 1400 Xg. This step was repeated at least four times. The cleaned worms were resuspended into 150 ml distilled water and passed through a 125 ml separatory funnel packed with glass microbeads (bead diameter about 0.5 mm). The first 10 to 20 ml of effluent contain exclusively L2 larvae (about 320 ± 30 μ (S.D.)); all other sizes are excluded. About 10,000 L2 larvae were added to each Petri plate containing 5 ml of clear 1% barley solution. The ensuing growth to maturation was highly synchronous. When adult worms were collected, care was taken to avoid gravid females. The synchronization technique with ensuing growth yields pure samples of all stages except the intrauterine L1 stage.

Nematode lengths were measured after the worms had been fixed and stained in 0.01% cotton blue lactophenol (Goodey, '57).

Extraction procedure

About 0.5 ml of synchronized nematodes were collected and washed three times by centrifugation with 0.05 M Tris-HCl buffer (pH 7.4). The bacterial count in the final wash fluid was negligible. The organisms were agitated with an equal volume of glass beads in a Mickle homogenizer for 15 minutes at 4°C. The homogenate was spun for four minutes at 450 xg and the pellet discarded. The protein concentration of the supernatant fluid was determined by the Lowry et al. ('51) method.

Analysis

Protein analysis was by polyacrylamide gel electrophoresis (Ornstein, '64). The electrophoresis procedures employed in preparing, staining, and destaining of the polyacrylamide gels, are described in the Canalco Instruction Manual (Canalco, Bethesda, Maryland). New batches of acrylamide and bisacrylamide were always recrystallized (Loening, '67). In every run, each gel was loaded with about 200 μg of protein. When polymerization of the sepa-

rating gel was catalyzed with ammonium persulfate or riboflavin and light the same results were obtained. Thus, there is no reason to suspect ammonium persulfate induced artifacts (Brewer, '67; Mitchell, '67).

Enzyme assays

Lactate dehydrogenase (LDH) was visualized by the method of Allen ('61). The malate dehydrogenase (MDH) staining mixture was the same as the LDH assay except malate replaced lactate. Esterase activity was marked with α-naphthyl acetate as the substrate (Markert and Hunter, '59). Alkaline phosphatase (Alk. P.) was analyzed by the lead conversion method (Allen and Hyncik, '63). Acid phosphatase (Acid P.) was demonstrated by a modified form of Gomori's method ('52). Gels were assayed for leucine aminopeptidase (LAP) with β-leucylnaphthylamide as substrate (Nachlis et al., '63).

All enzymatic determinations were made at least six separate times. No non-specific enzymatic reactions were observed in the gels when the substrate was omitted from each assay mixture.

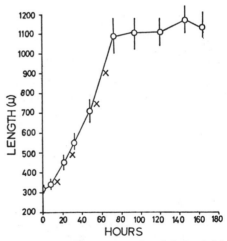

Fig. 1 Synchronous growth of isolated L2 larvae of *Panagrellus silusiae*. The vertical lines are 95% confidence limits. The X's denote the approximate mean length of the postpartum stages during development, i.e., L2, L3, L4 and adult stages.

The densitometric tracings of stained gels were made with a Joyce-Loebl Chromoscan.

RESULTS

Growth

The time course of growth of nematodes after isolation of L2 stage organisms is shown in figure 1. Although the curve represents one experimental run, the repeat experiments always show a similar pattern. Adults were never collected beyond 100 hours growth for protein analysis, but the additional measurements show that young adults do not differ significantly in length from the older and more mature worms. Second generation offspring are omitted from the curve. These young larvae start appearing at 120 hours. As the figure indicates after the isolation of L2 larvae if collections are made at 20 hour intervals pure samples of each developmental stage can be obtained.

Protein analysis

Aniline black stained gels of L2, L3, L4 and adult stages reveal no striking differences with respect to the number and positions of the bands (fig. 2). The juvenile

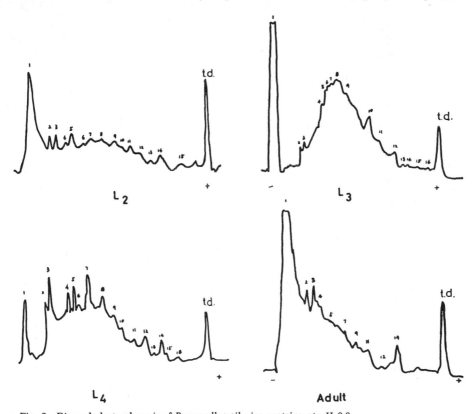

Fig. 2 Disc gel electrophoresis of *Panagrellus silusiae* proteins at pH 8.3.
Densitometric tracings of aniline black stained gel patterns of the postpartum stages; (a) L2 larvae, (b) L3 larvae, (c) L4 larvae and (d) adult stage. In this figure as well as in figures 3–7 the anode (+) is to the right and the cathode (−) on the left. T. D. marks the location of the tracking dye. The scanning was done on intact gels; peak heights are relative and the degree of background staining is not the same for all gels.

TABLE 1

Relative concentration of aniline black stained bands in polyacrylamide gels of L2, L3, L4 and adult stages of P. silusiae

Stage	Protein band															
	1	2	3	4	5	6	7	8	9	10	11	12	13	14	15	16
L2	++++	++	++	+	+++	+	+	+	++	+	++	+	+	++	+	±
L3	++++	++	++	+	+	++	+	+++	++	++	++	++	+	++	+	+
L4	+++	++	++	++	++	++	++	++	++	++	++	++	+	++	+	+
Ad.	++++	++	++	±	+++	±	+	−	++	+	++	+	−	++	+	−

The bands are numbered from the origin to the end of the gel. Designations of the intensity of the staining: ++++, heavy; +++, dark; ++, moderate; +, light; ±, barely visible; −, trace or no staining.

stages (L2, L3 and L4) have 16 distinguishable bands, whereas the adult stage homogenates usually show ten bands clearly with six additional but faint bands. The major difference between the protein patterns of the different stages is the relative concentration of certain bands. The extent of the relationships of the 16 bands of the different stages are catalogued in table 1 where the bands are numbered consecutively from the origin.

Enzyme analysis

In each electrophoretic analysis of a particular postembryonic stage 12 or 24 gels were run in parallel. Sets of two or four gels were analyzed for protein patterns with aniline black and for lactate dehydrogenase, malate dehydrogenase, α-naphthyl acetate esterase, alkaline phosphatase and acid phosphatase activity.

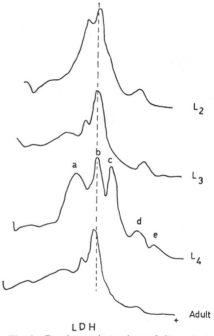

Fig. 3 Densitometric tracings of disc gel electrophoresis patterns of lactate dehydrogenase activity in L2, L3, L4 and adult stages of P. silusiae. The dotted line marks identical distances on the original gels in this figure as well as in figures 4 and 5.

(i) Lactate dehydrogenase (LDH). The L4 larvae show five discrete LDH bands (fig. 3). The bands are labelled a, b, c, d, and e from the origin. Bands c and e are L4-specific LDH enzymes and under normal incubation conditions are never observed in any other stage. Bands a, b and d are present in each of the other stages. The relative proportions of the LDH isozymes vary slightly from run to run; in general, band b predominates followed by bands a and then band d except in L4 larvae where band c is prevalent.

Comparing the locations of bands in aniline black stained gels with enzyme assayed gels has limitations since the aniline black stained bands include more than a single kind of protein. For example, band 14 is present in both L4 and adult stages. But, only the L4 extracts show LDH activity in this band.

When extracts of adult stages are incubated for 48 hours instead of one half hour, two very faint bands occupying sites identical to LDH-c and LDH-e of the L4 stage are present. Similar long incubations of extracts with stages prior to the L4 stage reveal no LDH-c or LDH-e activity.

(ii) Malate dehydrogenase (MDH). Each stage in the developmental sequence has a different number of MDH bands (fig. 4). The MDH-c band is present in all stages and MDH-e activity is found in both L3 and L4 larval stages. MDH-a, MDH-d and MDH-f bands are unique to L3 larvae, MDH-b is found exclusively in adult worms and MDII-g is found only in L4 larvae.

(iii) Esterase. The L4 larvae show six distinct bands with esterase activity. There is no difference among the three remaining stages with each having the same four esterase bands (fig. 5). When α-naphthyl butyrate was used as a substrate esterase bands c, d and e showed activity in gels of each of the stages. When the substrate was α-naphthyl laurate one faint band was

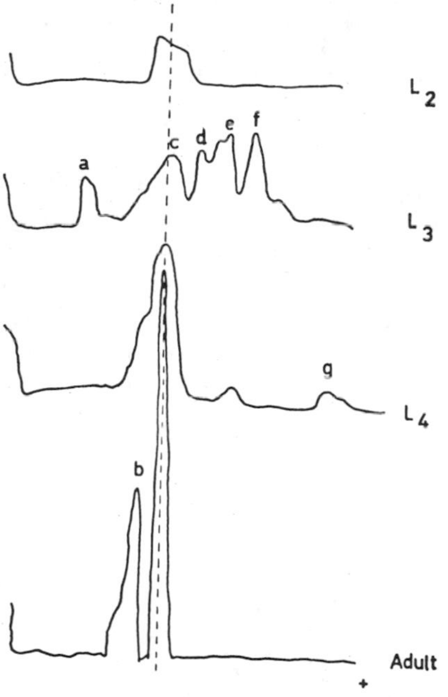

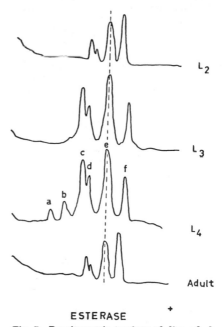

MDH

Fig. 4 Densitometric tracings of disc gel electrophoresis patterns of malate dehydrogenase activity in L2, L3, L4 and adult stages of *P. silusiae.*

ESTERASE

Fig. 5 Densitometric tracings of disc gel electrophoresis patterns of α-naphthyl acetate esterase activity in L2, L3, L4 and adult stages of *P. silusiae.*

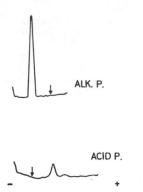

Fig. 6 Densitometric tracings of gels showing alkaline and acid phosphatase activity. The upper tracing of alkaline phosphatase activity is from an L2 larval extract. The lower tracing of acid phosphatase activity is from an L4 larval extract. The arrows indicate the position of the untested phosphatase on each gel.

observed in each stage at the location of esterase c or d or between them.

(iv) Acid phosphatase (A. P.). This enzyme is found as a single band at the same gel location in each stage (fig. 6).

(v) Alkaline phosphatase (Alk. P.). This enzyme is present as a single protein band at the same site in gels of each stage (fig. 6).

(vi) Leucine aminopeptidase. Although leucine aminopeptidase is prevalent during certain stages in the life cycle of parasitic nematodes (Rogers, '65), it was never detected in any homogenates of *P. silusiae.*

The results of the electrophoretic analyses with respect to enzyme activity and stage specificity are summarized in figure 7.

DISCUSSION

The present study demonstrates clearly that each free-swimming stage during the growth cycle of *P. silusiae* has a unique pattern of enzymic proteins. It should be noted that these changes take place when somatic cell division is supposedly suppressed and in the absence of tissue differentiation at least until the L4 stage when ovaries and testes begin to develop. At this time, we cannot assign any of the observed enzymes to a specific tissue location, the high degree of enzyme heterogeneity of L4 larvae may be due in part to gonad specific proteins.

No studies similar to the one reported here have been conducted on free-living nematodes. Bloom and Entner ('65), however, have shown that in the parasitic nematode *Ascaris lumbricoides* the enzymatic properties of larval MDH differ from those of adult MDH. This result suggests that *Ascaris* has at least two MDH isozymes that are stage-specific. Benton and Myers ('66) investigated α-naphthyl acetate esterase activity in homogenates of *Panagrellus redivivus* containing unknown proportions of all stages. At pH 8.3 in polyacrylamide gel they found six esterase specific bands. We have found that nonstage-specific extracts of *P. redivivus* usually reveal four esterase bands at pH 8.3 (unpublished results). These esterases have similar migratory rates as bands a, c, d and e in *P. silusiae.* Benton and Myers

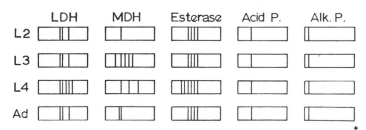

Fig. 7 A schematic representation of acrylamide gels depicting the enzymes found in different stages of *P. silusiae* during postembryonic development. The thickness of the bands does not represent concentration levels in the original gels nor is the spacing of the bands to scale. Comparable bands are kept in line. LDH, lactate dehydrogenase; MDH, malate dehydrogenase; Acid P, acid phosphatase; Alk. P. alkaline phosphatase; Esterase, α-naphthyl acetate esterase.

also report that nonstage-specific homogenates of *P. redivivus* have two bands with acid phosphatase activity and no alkaline phosphatase activity whatsoever at pH 8.3. *P. silusiae*, on the other hand, has one of each kind of phosphatase present without qualitative variation throughout the life cycle.

We have observed with *P. redivivus* at least two bands with alkaline phosphatase activity and at least the same number of bands with acid phosphatase activity (unpublished results).

We cannot fully explain the differences between the results recorded by Benton and Myers for *P. redivivus* and ours with this species; although the disparity with regard to alkaline phosphatase activity may be due to the different techniques used. We employed the lead conversion method whereas they utilized a dye coupling method. But the differences observed between the two species, *P. redivivus* and *P. silusiae*, are of major importance since one of us (JP) has evidence which indicates that these two worms will readily breed and give fertile offspring. The opportunity is available to analyze the genetic basis for the observed protein differences.

The sequential changes of stage specific proteins suggest that a regulatory program is operating during postembryonic development. We have no information to indicate that the periodic appearance of enzymes is determined at the gene level. It is evident that the presence of activity of an enzyme is not necessarily equivalent to enzyme biosynthesis. This means that the degree of gene activity cannot be gauged unless more direct assays are employed (Ursprung et al., '68). These experiments are in progress.

ACKNOWLEDGMENT

This work has been submitted by Hsiao-mei Hu Chow in partial fulfillment of the requirements for an M.Sc. degree at the University of Waterloo.

This investigation was supported by grants A-3491 from the National Research Council of Canada and the Brown-Hazen Fund. H. H. C. is the recipient of Province of Ontario Research Fellowship.

LITERATURE CITED

Allen, J. M. 1961 Multiple forms of lactic dehydrogenase in tissues of the mouse: Their specificity, cellular localization, and response to altered physiological conditions. Ann. N. Y. Acad. Sci., 94: 937–951.

Allen, J. M., and C. Hyncik 1963 Localization of alkaline phosphatases in gel matrices following electrophoresis. J. Histochem. Cytochem., 11: 169–175.

Benton, A. W., and R. F. Myers 1966 Esterases, phosphatases and protein patterns of *Ditylenchus triformis* and *Panagrellus redivivus*. Nematologica, 12: 495–500.

Bloom, S., and N. Entner 1965 Mitochondrial enzymes in developing larvae of *Ascaris lumbricoides*. Biochem. Biophys. Acta, 99: 22–31.

Brewer, J. M. 1967 Artifact produced in disc electrophoresis by ammonium persulfate. Science, 156: 256–257.

Chitwood, B. G. 1950 Introduction to Nematology. B. G., and M. B. Chitwood, eds. Monumental Printing Co., Baltimore, Md.

Goodey, J. B. 1957 Laboratory methods for work with plant and soil nematodes. Technical Bulletin No. 2. Her Majesty's Stationery Office, London.

Gomori, G. 1952 Microscopic Histochemistry. Univ. Chicago Press, Chicago.

Gysels, H., and W. van der Haegen 1962 Postembryonale ontwikkeling en vervellingen van de vrijlevende Nematode *Panagrellus silusiae* (de Man, 1913), Goodey, 1945. Natuurwet. Tijdschr., 44: 3–20.

Loening, U. E. 1967 The fractionation of high-molecular-weight ribonucleic acid by polyacrylamide-gel electrophoresis. Biochem. J., 102: 251–257.

Lowry, O. H., N. J. Rosebrough, A. L. Farr and R. J. Randall 1951 Protein measurement with the Folin phenol reagent. J. Biol. Chem., *193:* 265–275.

Markert, C. L., and R. L. Hunter 1959 The distribution of esterases in mouse tissues. J. Histochem. Cytochem., *7:* 42–49.

Mitchell, W. M. 1967 A potential source of electrophoretic artifacts in polyacrylamide gels. Biochem. Biophys. Acta, *147:* 171–174.

Nachlas, M. M., B. Monis, D. Rosenblatt and A. M. Seligman 1960 Improvement in the histochemical localization of leucine aminopeptidase with a new substrate, L-leucyl-4-methoxy-2-naphthylamide. J. Biophys. Biochem. Cytol., *7:* 261–264.

Ornstein, L. 1964 Disc electrophoresis. I. Background and Theory. Ann. N. Y. Acad. Sciences, *121:* 321–349.

Rogers, W. P. 1965 The role of leucine aminopeptidase in the moulting of nematode parasites. Comp. Biochem. Physiol., *14:* 311–321.

Samoiloff, M. R., and J. Pasternak 1968 Nematode Morphogenesis: Fine structure of the cuticle of each stage of the nematode, *Panagrellus silusiae* (de Man, 1913), Goodey, 1945. Can. J. Zool., *46:* 1019–1022.

Ursprung, H., K. D. Smith, W. H. Sofer and D. T. Sullivan 1968 Assay systems for the study of gene function. Science, *160:* 1075–1081.

THE FUNCTION OF LEUCINE AMINOPEPTIDASE IN EXSHEATHING FLUID

W. P. Rogers

Rogers (1963, 1965) obtained results which suggested that leucine aminopeptidase might be responsible for the initial breakdown of sheaths during exsheathment. Two species of infective juveniles, *Haemonchus contortus* and *Trichostrongylus colubriformis*, were used. Recently Ozerol and Silverman (1969) got evidence from which they concluded "that leucine aminopeptidase is *not* the enzyme responsible for the exsheathment of *H. contortus.*" This view was based on the findings that (a) the factor which attacked isolated sheaths was heat-stable, and (b) leucine aminopeptidase was absent from their preparations of exsheathing fluid.

The reasons why Ozerol and Silverman failed to find leucine aminopeptidase in their preparations of exsheathing fluid are obvious (see Discussion). Their finding that the isolated sheaths may be attacked by a heat-stable substance does not necessarily mean that the system is important in vivo. Indeed their method of preparing isolated sheaths, and the observation in control experiments that Earle's solution alone showed 35% activity, suggest that the sheaths were not normal. Though more than 100 isolated sheaths have been used in experiments described in the present paper spontaneous "ring formation" was never seen and Earle's solution was inactive. It seems, therefore, that the most fruitful way of resolving this controversy would be to provide simple methods which could be widely used, even in class experiments, to show that exsheathing fluid which contains leucine aminopeptidase also contains a heat-labile factor which attacks sheaths.

It is difficult to obtain active exsheathing fluid because it is secreted in small amounts. Moreover, if leucine aminopeptidase is the exsheathing enzyme, there is the added difficulty of stimulating the juveniles and collecting the exsheathing fluid under conditions which will not inactivate this highly labile enzyme (Smith and Hill, 1960). In earlier work (Rogers and Sommerville, 1960; Rogers, 1963, 1965) efforts were made to keep as close as possible to physiological conditions. Thus exsheathing fluid was obtained using natural systems in vivo or by using reducing conditions and bicarbonate–40% carbon dioxide as a buffer near neutrality to simulate conditions in the rumen.

In the present work emphasis has been placed on finding simple methods for obtaining active exsheathing fluid and no special care to use physiological conditions was taken. Thus the reducing agent was omitted from the stimulating medium and 100% carbon dioxide was used as the gas phase with tris (hydroxymethyl) aminomethane as the buffer. The same system was used for collecting the exsheathing fluid after the carbon dioxide had been removed.

There seems no reason why the micro methods which are already available for the estimation of leucine aminopeptidase (Rogers, 1964) should not be used. However, Ozerol and Silverman (1969) chose a method, used in clinical investigations for the estimation of leucine aminopeptidase in urine and serum, which is both insensitive and unspecific. For this reason a simplification of a previously described method is given.

Although it was stated (Rogers, 1965) that freeze-drying causes losses in the activity of the leucine aminopeptidase in exsheathing fluid, Ozerol and Silverman (1969) used this method in their attempts to concentrate the enzyme. In this paper simple, quick methods for doing this are therefore described.

Because fresh sheaths, which have had the minimum exposure to unphysiological conditions, may be required for the demonstration of the normal exsheathing action, the method of preparing isolated sheaths which was used previously (Rogers, 1965) is described in detail.

MATERIALS AND METHODS

Infective juveniles of *H. contortus* were used for all experiments. They were stored, for periods up to 1 month, in water at 5 C until they were needed. Contamination was reduced by frequent washing with water.

Preparation of exsheathing fluid

About 2 to 3 $\times$ 10^6 juveniles which had been washed by sedimentation in water were washed once in 0.15 M tris-hydrochloric acid buffer containing 10^{-3} M Mg^{++} at pH 8.3. The juveniles were then suspended in the buffer to give a total volume of 8 ml. The suspension was gassed for 5 min with 100% carbon dioxide. This reduced the pH to about 6.7. The juveniles were then incubated under 100% carbon dioxide for 28 min in a slowly oscillating water bath at 37.6 C. Warm, water-saturated, carbon dioxide-free air was passed for 5 min through the suspension which was held in a bath at 37.6 C. During this treatment the pH rose to about 7.8. From this point samples were taken at intervals to estimate the degree of exsheathment which had occurred. Under normal circumstances the first sample showed less than 1% exsheathment. After 60 min, when exsheathment had reached about 75%, the diluted exsheathing fluid, 5 to 7 ml, was collected in a cold container by filtration through sintered glass of medium porosity.

Concentrating the exsheathing fluid

Two methods were used.
(a) With carbowax 20M. Pale yellow, dilute exsheathing fluid was placed in dialysis tubing 1 cm in diameter. The sac was placed in powdered carbowax at 5 C for several hours until the volume was reduced to about 0.5 ml of reddish solution. This was washed from the sac with 0.2 ml of cold 0.15 M tris-Mg^{++} buffer at pH 8.3.
(b) With Sephadex. Dilute exsheathing fluid was mixed with Sephadex G-25 (1 ml/0.36 g) in a cylindrical column, 1 cm diameter, fitted with a sintered glass filter disk of medium porosity. The mixing was carried out a little at a time so that the penetration of the liquid through the dry powder was complete. After 10 min the concentrate was collected, by suction, in a cold container. (A second, less concentrated fraction, could be obtained by washing the column of Sephadex with tris-Mg^{++} buffer, 0.1 ml/0.36 g of dry Sephadex.)

This procedure (Flodin et al., 1960) provided a clear, buffered concentrate of the macromolecular fraction, about 1/10th of the original volume.

Preparation of isolated sheaths

Fresh, active juveniles were cut with a fine stainless steel knife about $\frac{1}{3}$ the length from the anterior end. After 10 min the cut sheaths had lost their contents and the anterior fragments could be transferred to test solutions. This was most easily done with a fine glass seeker with a small knob on the end. With a slow rotating movement the selected part of a sheath was propelled to the surface where it was held by surface tension. It was removed from the surface by a sharp lifting motion of the seeker and transferred to the test solution. The microdissection was carried out under a dissecting microscope at the lowest power at which the isolated sheaths could conveniently be seen.

This procedure is considered to be an important part of the technique for testing solutions for exsheathing activity. In contrast to the method of Ozerol and Silverman (1968), it made it possible to have sheaths in a test solution within 15 min of dissection and without prior exposure to solutions other than distilled water.

Tests with isolated sheaths

Test solutions, 10 μl in volume, were placed on slides within rings of vaseline. Anterior ends of sheaths, usually 5 for each test, were placed in the solutions and cover slips were sealed in position with hot wax. Preparations were incubated at 38 C and examined, at intervals up to 24 hr for control solutions, for "ring formation" (Sommerville, 1957). The activity of concentrated exsheathing fluid could usually be detected within 20 min.

Tests for leucine aminopeptidase

L-leucinamide hydrochloride (British Drug Houses) contained relatively small amounts of ammonia and gave moderate blanks. Other commercial products which were tested required repeated recrystallization before use. The method of Smith and Slonin (1948) for preparing this substrate was most satisfactory and gave a product containing traces only of ammonia (Rogers, 1964).

The estimation of enzyme activity was made in standard Conway vessels. In the outer ring was placed 0.25 ml of concentrated enzyme solution and 5 μl of 1 M leucinamide in 0.15 M tris-Mg^{++} adjusted to pH 8.3. Sulphuric acid, 0.2 ml of 0.01 N, was placed in the inner region. The sealed vessels were incubated at 38 C for 2 hr, then 0.5 ml of saturated potassium carbonate was added to the outer ring and distillation allowed to take place for 2 hr. After adding 50 μl of Nessler solution to the acid, NH$_3$/N was measured as the optical density in a Beckman–Spinco microspectrocolorimeter at 420 mμ.

Enzyme activity was also measured spectrophotometrically (Mitz and Schleuter, 1958). The experimental cell, a quartz microcuvette, contained 0.4 ml of 0.05 M L-leucinamide hydrochloride in

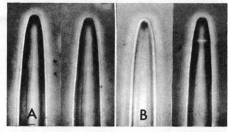

FIGURE 1. The effect of heated (A) and unheated (B) concentrated exsheathing fluid on isolated sheaths. A and B, left, at time zero; A and B, right, after 10 min. Twenty-four hours later the control was still unchanged.

0.15 tris-Mg^{++} buffer at pH 8.3 and 0.2 ml of concentrated exsheathing fluid. The control cell contained 0.4 ml 0.05 M L-leucine in tris-Mg^{++} buffer and 0.2 ml of the exsheathing fluid. The breakdown of the substrate at 38 C was followed at 238 mμ.

RESULTS

With isolated sheaths as substrates

"Ring formation" occurred with highly concentrated preparations of exsheathing fluid, within 20 min at 38 C. (In Figure 1 the change in the sheath has gone beyond "ring formation" in 10 min.) Such results were obtained with material which had been concentrated by exclusion from Sephadex G-25 as well as that concentrated by dialysis, i.e., the activity lay in the macromolecular fraction of exsheathing fluid. Activity was not found even after 24-hr incubation in any sample of concentrated exsheathing fluid which had been heated at 80 C for 15 min (Fig. 1).

The activity of the exsheathing fluid was

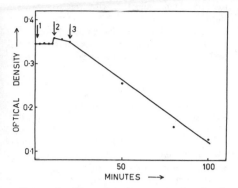

FIGURE 2. The action of concentrated exsheathing fluid on L-leucinamide measured at 238 mμ. 1, substrate only, 20 C; 2, exsheathing fluid added, 20 C; 3, incubation at 38 C commenced.

completely inhibited by 10^{-2} M diaminoethanetetra-acetic acid, Cu^{++} or Hg^{++}.

With L-leucinamide as substrate

The activity of leucine aminopeptidase was more difficult to detect than exsheathing activity. Dilute solutions of exsheathing fluid would often show some action on isolated sheaths after 3 hr at 38 C but would not hydrolyze sufficient leucinamide to give a result appreciably above the level of controls. It should be pointed out, however, that the visual detection of activity on sheaths at high magnification would be a method of great sensitivity.

Results obtained in two experiments in which ammonia released from L-leucinamide was used as a measure of enzyme activity are shown in Table I. Ammonia excreted by the

TABLE I. *The activity of leucine aminopeptidase in concentrated exsheathing fluid. In (a) the substrate was purified by recrystallization; in (b) a commercial product was used.*

Reaction mixture	Incubation time (min at 38 C)	NH₃/N μg found		NH₃/N μg/hr/ml exsheathing fluid	
		(a)	(b)	(a)	(b)
0.25 ml exsheathing fluid 5 μl 1 M leucinamide	0	2.5	5.2		
0.25 ml exsheathing fluid 5 μl 1 M leucinamide	120	7.6	10.0		
				10.2	9.6
0.25 ml tris-Mg^{++} 5 μl 1 M leucinamide	0	0.4	3.3		
0.25 ml tris-Mg^{++} 5 μl 1 M leucinamide	120	0.4	3.4		

juveniles and collected with the exsheathing fluid gave the high values in the controls.

Figure 2 shows results obtained when the breakdown of leucinamide by concentrated exsheathing fluid was measured spectrophotometrically.

DISCUSSION

The methods described above have the advantages that they are simple and specific. However, they have disadvantages as well. The use of 100% carbon dioxide for stimulation may have led to the occurrence of a yellow pigment in exsheathing fluid because it was not found when 40% carbon dioxide was used. It is possible then that unphysiological conditions led to increased leakage or secretion of material from the juveniles. The estimation of activity of leucine aminopeptidase by the release of ammonia from leucinamide had the advantage of sensitivity and specificity (Smith and Slonin, 1948) but the method required a highly purified substrate. Moreover, ammonia is excreted in the course of metabolism of H. contortus and so it was present in exsheathing fluid which thus gave high "blank" values in control experiments. This ammonia could be removed by dialysis. However, the qualitative demonstration of the presence of the enzyme was sufficient for the purpose of this investigation so that the ammonia was removed only in some preliminary experiments.

The period of time for which juveniles were stimulated was critical. If the period was too long, exsheathment commenced before carbon dioxide was removed from the system, and the low pH led to destruction of enzyme. Too short a period caused low exsheathment and a poor yield of enzyme. The temperature at which stimulation was conducted was also critical. Changes of less than 0.5 C in the region of 38 C had a marked effect on the time needed for optimal stimulation.

Though the chief aim of this work was the simplification of methods, some additional evidence, e.g., the macromolecular nature of the exsheathing substance and its inhibition by Cu^{++} and Hg^{++}, which links exsheathing substance and leucine aminopeptidase, has been found. It is suggested (see below) that there is already reasonable evidence supporting the view that leucine aminopeptidase is the exsheathing enzyme. A more rigorous test would require purification of the enzyme in exsheathing fluid, a difficult task because (a) the amounts available are small, (b) the enzyme is unusually unstable (Smith and Hill, 1960), and (c) it tends to aggregate during purification. A histochemical examination of the parts of sheaths normally attacked by the exsheathing substance would also help.

Evidence supporting the view that leucine aminopeptidase is the exsheathing substance

(1) Exsheathing fluid from H. contortus obtained under physiological conditions, i.e., in dialysis sacs in a rumen fistula, required Mn^{++} or Mg^{++} as a cofactor for its action on isolated sheaths (Rogers and Sommerville, 1960). Similarly, exsheathing fluid from H. contortus and Trichostrongylus colubriformis obtained in vitro required one or other of these cofactors for its action on isolated sheaths and for the hydrolysis of L-leucinamide. Moreover, 0.01 M diaminoethanetetra-acetic acid inhibited both the enzyme and the exsheathing factor (Rogers, 1965). The present work has shown that Cu^{++} and Hg^{++}, which are inhibitors of leucine aminopeptidase (Smith and Hill, 1960), also inhibit the exsheathing fluid of H. contortus.

(2) Heating for 15 min at 80 C destroyed the activity of exsheathing fluid. The "turning point" was about 42 C (Rogers and Sommerville, 1960).

(3) A variety of conditions which destroyed exsheathing activity, e.g., pH levels below 7 and the absence of cofactor during purification, also inactivated the enzyme (Rogers, 1963, 1965; Smith and Hill, 1960).

(4) Leucine aminopeptidase was found in exsheathing fluid of H. contortus collected at a time when exsheathing occurred. It was not secreted before exsheathment nor for at least several hours after exsheathment was complete. No enzyme was released by unstimulated juveniles and large numbers of experiments have shown that most active preparations have been obtained when exsheathment was high and vice versa. Juveniles which had been exsheathed artificially with sodium hypochlorite did not secrete the enzyme when incubated without stimulation and only small amounts of enzyme were released when juve-

niles were stimulated a second or third time (Rogers, 1963, 1965).

(5) Fractionation of exsheathing fluid on acrylamide columns showed that only those fractions which had leucine aminopeptidase activity were active against isolated sheaths. The active fractions contained less than 1% of the total protein in the exsheathing fluid used for fractionation. However, it was necessary to incubate samples taken from the column for 12 hr at 38 C before carrying out tests with isolated sheaths. Thereafter those samples which contained leucine aminopeptidase were found to attack sheaths in the normal way. As there is evidence that mammalian leucine aminopeptidase may aggregate under certain conditions (Spackman et al., 1955) these results were explained on the basis of the aggregation and disaggregation of the enzyme (Rogers, 1965).

(6) Methods used for concentrating the exsheathing substance also concentrated leucine aminopeptidase. Thus concentrating the macromolecular fraction of exsheathing fluid by exclusion from Sephadex G-25 (Flodin et al., 1960) decreased the time needed for "ring formation" from 4 hr to 5 min and increased the activity of leucine aminopeptidase from undetectable to a level which hydrolyzed 10.2 μg L-leucinamide per hr per ml.

(7) Strong evidence that leucine aminopeptidase is the molting enzyme has been obtained by Davey and Kan (1967) who stated: "The neurosecretory cycle which occurs during the last molt of *Phocanema decipiens* cultured *in vitro* closely parallels the cycle of synthesis and release of leucine aminopeptidase by the excretory gland. Culturing the worms in non-nutrient saline permits the deposition of a new cuticle, but prevents ecdysis. Under these conditions, the neurosecretory cycle, and the synthesis of the enzyme are inhibited. Saline extracts of the anterior ends of worms prepared at the time of maximum neurosecretion bring about the synthesis of leucine aminopeptidase by the excretory glands *in vitro*."

(8) Histochemical tests with *Xiphinema index* (Roggen et al., 1967) showed that leucine aminopeptidase was present in vacuoles in the epidermis. During the intermolt period the number of these vacuoles increased to reach a maximum just before the molt occurred. During the molt most of the enzyme-containing vacuoles disappeared and at the end of the molt the leucine aminopeptidase could be detected in the fluid between the old and the new cuticle.

Evidence against the hypothesis

(1) Ozerol and Silverman (1969) found that the substance in their preparations of exsheathing fluid which attacked isolated sheaths was heat stable. It should be emphasized, however, that sheaths themselves are not highly stable structures. Thus they will dissolve completely in boiling water and in 0.2 N sodium hydroxide (Bird and Rogers, 1956). There may be many heat-stable substances which cause changes in the sheath similar to those which occur under physiological conditions.

(2) Leucine aminopeptidase was not found in exsheathing fluid prepared by Ozerol and Silverman (1969). This was probably due to a combination of the following factors: (a) the enzyme was destroyed during the collection and concentration of exsheathing fluid (pH about 6.1, incubation for 24 hr at 38 C, freeze-drying), (b) the insensitivity of the method used for detecting the enzyme, and (c) the use of smaller numbers of juveniles for the preparation of exsheathing fluid.

(3) It could be argued that both the leucine aminopeptidase and the heat-labile exsheathing substance were produced by contaminating organisms. This seems unlikely, however, because both these substances were produced only from juveniles which were exsheathing.

(4) The temporary loss of exsheathing activity in preparations which had been fractionated on acrylamide columns has been attributed to the aggregation of the enzyme. However, the "aggregated enzyme" did not lose its activity against low molecular weight substrates. To explain this Rogers (1965) suggested that aggregation would be more likely to affect the action of the enzyme on a substrate which is part of an organized insoluble structure rather than on a soluble substrate of low molecular weight.

(5) Mammalian leucine aminopeptidase does not attack isolated sheaths. The enzymes in exsheathing fluid, however, show

35

a high specificity. Thus, like mammalian leucine aminopeptidase, the enzymes in exsheathing fluid of *H. contortus* and *T. colubriformis* hydrolyze L-leucinamide and L-leucyl-β-naphthylamide (though the affinity of the enzymes for the two substrates varies considerably) but they are entirely specific to their natural substrates (Rogers, 1965).

ACKNOWLEDGMENTS

I wish to thank Mrs. M. Ross for teaching me her technique of microdissection, and Mr. N. Stewart who did the photography. The juveniles used in these experiments were kindly provided by CSIRO McMaster Laboratory.

This work was supported by a Public Health Service Grant AI04093-07 for which I am most grateful.

LITERATURE CITED

BIRD, A. F., AND W. P. ROGERS. 1956. Chemical composition of the cuticle of third stage nematode larvae. Exp. Parasit. **5**: 449–457.

DAVEY, K. G., AND SAN PHENG KAN. 1967. An endocrine basis for ecdysis in a parasitic nematode. Nature, London **214**: 737–738.

FLODIN, P., B. GELOTTE, AND J. PORATH. 1960. A method for concentrating solutes of high molecular weight. Nature, London **188**: 493–494.

MITZ, M. A., AND R. J. SCHLUETER. 1958. Direct spectrophotometric measurement of the peptide bond: application to the determination of acylase 1. Biochim. Biophys. Acta **27**: 168–172.

OZEROL, N. H., AND P. H. SILVERMAN. 1969. Partial characterization of *Haemonchus contortus* exsheathing fluid. J. Parasit. **55**: 79–87.

ROGERS, W. P. 1963. Physiology of infection with nematodes: some effects of the host stimulus on infective stages. Ann. N. Y. Acad. Sci. **113**: 208–216.

———. 1964. Micromethods for the study of leucine aminopeptidase. Microchem. J. **8**: 194–202.

———. 1965. The role of leucine aminopeptidase in the moulting of nematode parasites. Comp. Biochem. Physiol. **14**: 311–321.

———, AND R. I. SOMMERVILLE. 1960. The physiology of the second ecdysis of parasitic nematodes. Parasitology **50**: 329–348.

ROGGEN, D. R., D. J. RASKI, AND N. O. JONES. 1967. Further electron microscopic observations of *Xiphinema index*. Nematologica **13**: 1–16.

SMITH, E. L., AND R. L. HILL. 1960. Leucine aminopeptidase. *In* P. D. Boyer, H. Lardy, and K. Myrbäck (eds.), The Enzymes **4**: 37–60. Academic Press, New York.

———, AND N. B. SLONIN. 1948. The specificity of leucine aminopeptidase. J. Biol. Chem. **176**: 835–841.

SOMMERVILLE, R. I. 1957. The exsheathing mechanism of nematode infective larvae. Exp. Parasit. **6**: 18–30.

SPACKMAN, D. H., E. L. SMITH, AND D. M. BROWN. 1955. Leucine aminopeptidase. IV. Isolation and properties of the enzyme from swine kidney. J. Biol. Chem. **212**: 255–269.

NEMATODE BIOCHEMISTRY—X. EXCRETION OF GLYCEROL BY FREE-LIVING NEMATODES

MORTON ROTHSTEIN

INTRODUCTION

DURING the course of previous investigations, it was found that the small, free-living nematode, *Caenorhabditis briggsae*, during incubation with various radioactive substrates, produced in the medium a large variety of labeled products (Rothstein, 1963; Rothstein & Mayoh, 1964; Rothstein, 1965). In general, the greatest amount of radioactivity was found in the respective amino acids most closely related to the substrate. Thus, incubation with acetate-2-^{14}C resulted in the accumulation in the medium of highly labeled glutamate, aspartate and alanine; pyruvate-1-^{14}C yielded labeled alanine and aspartate; formate-^{14}C yielded labeled serine, and aspartate-4-^{14}C yielded labeled glycine (via the action of isocitrate lyase).

In addition to amino acids, substantial amounts of radioactive acidic and neutral materials were produced from labeled acetate (Rothstein, 1963). The present report deals with the latter group of products. They consist mainly of glucose, trehalose and glycerol. Of these, glycerol is the major radioactive product if *C. briggsae* is incubated with acetate-2-^{14}C in "whole medium". In sharp contrast, very little glycerol is formed if the incubation is carried out in water. These results and those obtained with two other free-living nematodes, *Turbatrix aceti* and *Panagrellus redivivus*, are discussed.

MATERIALS AND METHODS

In general, the nematodes were grown in axenic culture as previously described (Tomlinson & Rothstein, 1962; Rothstein & Cook, 1966), and sterility checks performed where necessary.

Neutral fractions

The following procedure is typical of all of the methods of isolation, although the products retained on Dowex-1-acetate were not examined in all experiments, nor were all experiments necessarily paired in "water" and "whole medium".

Comparison of products from incubations in "whole medium" and in water

Twelve-day cultures of *C. briggsae* grown in ten Erlenmeyer flasks (50 ml), were divided into two groups. The worms in six flasks were treated with sterile trypsin (Tomlinson & Rothstein, 1962); 24 hr later they were centrifuged and washed three times in the usual manner. The washed worms were divided between two 50-ml Erlenmeyer flasks each containing 4 ml of sterile water and 10 μc (20 μmoles) of sterile sodium acetate-2-^{14}C. At the same time, 10 μc (20 μmoles) of acetate-2-^{14}C were added to each of the four flasks of worms remaining in whole medium. After $3\frac{1}{2}$ days a sample from each flask was incubated with Difco "Brain Heart Infusion" as a sterility check. The worms were removed by centrifugation and washed two times with water. The media and washes from the "water" experiment were combined and the "whole medium" treated similarly. The latter was deproteinized by treatment with perchloric acid and subsequent neutralization with KOH. Both sets of media were then acidified and evaporated to near dryness on a rotary evaporator. The following procedure was applied to both media.

The concentrated medium was applied to a column of Dowex-1-acetate (200 × 10 mm) and the column treated with 50 ml of water (or until no further counts came off the column). In some cases, 20γ of carrier glycerol was added to the eluate which was then concentrated *in vacuo*, and passed through a column of Dowex-50 (H$^+$) (120 × 10 mm) with 50 ml of water. After evaporation of the water, samples of the residue were counted and chromatographed in BAW* (Fig. 1) and EAW, using glycerol, glucose and trehalose as standards. The remainder of the product from the "whole medium" experiment was streaked on Whatman No. 1 paper and run in BAW. By means of radioautography or strip scanning, the ^{14}C activity was located. Products were eluted and rechromatographed in other solvents (e.g. Fig. 2C).

From the original Dowex-1-acetate column, glutamic and aspartic acids were removed together by addition of 40 ml of 1 N acetic acid. This fraction was counted and the identity of the amino acids checked by paper chromatography.

Worm extracts

The worms from the respective experiments described above were homogenized in 15 ml of 80% ethanol containing 8 ml of glass beads† for 4 min at a setting of 80 V using a Sorvall Omni-Mixer. After removal of the beads and cell debris by centrifugation, 20γ of glycerol and 5 μmoles each of glutamate and aspartate were added to the respective extracts. The extracts were then concentrated and chromatographed on a column of Dowex-1-acetate (350 × 10 mm); acetic acid (0·5 N) was used for development (Hirs *et al.*, 1954). Ninhydrin color and radioactivity were determined for each fraction. After elution of the aspartic acid, 1 N HCl was used to remove all organic acids from the column.

The respective water eluates from the columns of Dowex-1-acetate were passed through a column of Dowex-50, and the eluates counted for radioactivity and treated on paper as above. From the extract of worms incubated in "whole medium", the amount of radioactivity due to glycerol was estimated from the relative size of the peaks after scanning of paper chromatograms (Fig. 2A).

* Abbreviations used: BAW, butanol–acetic acid–water, 4 : 1 : 1; EAW, ethanol–ammonia–water, 18 : 1 : 1; BPW, butanol–pyridine–water, 1 : 1 : 1; IBW, isopropanol–*n*-butanol–water, 7 : 1 : 2.

† Superbrite glass beads, 0·008 in. diameter, Minnesota Mining and Manufacturing Co. Before use, the beads were refluxed for 5 hr in 6 N HCl and then washed free of acid.

Soy-peptone medium

In a large number of experiments, cultures of nematodes were washed and incubated with acetate-2-^{14}C in fresh soy-peptone medium (Tomlinson & Rothstein, 1962) without the addition of liver extract. Incubations of this type were also used for the experiments with *T. aceti* and *P. redivivus*. Typically, the time of incubation was 3 days. The "whole medium" incubations with glycerol-^{14}C were performed in this manner (Fig. 5).

Other media components

Fourteen-day cultures of *C. briggsae* were treated with trypsin, washed and dispensed into single flasks containing only the respective components of the nematode basal medium (Sayre *et al.*, 1963) as follows: structural compounds, nucleotides, soy-peptone, vitamins, glucose, salts, glutathione, 0·1 M buffer, 0·07 M buffer, 0·03 M buffer. Acetate-2-^{14}C (10 μc; 5 μmoles) was added to each flask. After 3 days, each medium was passed successively through Dowex-1-acetate and Dowex-50 (H$^+$), concentrated, and 10 per cent of each product chromatographed on paper (BAW) and scanned for ^{14}C-activity.

Antibiotics

In some experiments, 100 units of penicillin and 100γ of streptomycin/ml were added to the incubation media. No differences in results were observed between experiments with and without added antibiotics.

Identification of products

Paper chromatography. BAW and EAW were used routinely for development of chromatograms. BPW or IBW was used where a third solvent system was deemed necessary. Glycerol, glucose and trehalose were detected by spraying with ammoniacal AgNO$_3$. Generally, 1–2·5 per cent of the total sample was applied to the paper.

Glycerol tribenzoate. A sample of presumed glycerol (99,000 counts/min) isolated from a "whole medium" experiment was mixed with 400 mg of carrier glycerol and the product converted to the tribenzoate (Abraham & Hassid, 1957). The product was recrystallized repeatedly from methanol. Samples weighing 7 mg were counted after each crystallization in a Nuclear-Chicago Scintillation counter. All samples were counted to 20,000 counts.

Dihydroxyacetone. A solution containing 12,000 counts/min of the neutral product (glycerol) from the above experiment was added to a few mg of dihydroxyacetone in water. A few ml of a saturated solution of 2,4-dinitrophenylhydrazine in 2 N HCl was added and the resulting reddish precipitate was filtered, recrystallized from ethanol–water and examined for radioactivity after plating on an aluminum planchet. The filtrate was extracted repeatedly with ethyl acetate and the extracts counted.

Conversion of trehalose to glucose

A sample of material running in the trehalose position (BAW) was eluted, rechromatographed on paper in IBW and eluted a second time. The product was hydrolyzed in 1·5 N H$_2$SO$_4$ for 12 hr, the sulfate removed by passage through Dowex-1-acetate, and the water eluate dried and chromatographed in BAW.

Degradation of glycerol

Experimentally derived samples of radioactive glycerol, after isolation from paper, were mixed with 0·5 m-moles of carrier glycerol (46 mg). The specific activity of the sample was determined based on this weight (the weight of isolated glycerol is negligible). After treatment with periodate, the resulting formaldehyde was distilled into a chilled receiver, and a solution of dimedone (400 mg) in 5 ml of 50% ethanol was added. After standing at room temperature for 30 min, the mixture was warmed briefly, chilled and the product filtered and recrystallized (usually twice) to constant specific activity (see Table 2).

Identification of glycerol

As a typical example, the radioactive product in the neutral peak from experiment 4 M (Table 1), matched exactly the position of glycerol in EAW, BAW and BPW. Since glycerol is not satisfactorily separated from dihydroxyacetone in the solvent systems used, the experiment with the latter was carried out as described above. The precipitated 2,4-dinitrophenylhydrazone of dihydroxyacetone contained no radioactivity, and ethyl acetate did not extract any counts from the aqueous layer: i.e. no other hydrazone was formed.

The tribenzoyl derivate of isolated glycerol to which carrier glycerol had been added had a specific activity of 383, 388 and 381 counts/min per 7·0 mg in crystallizations 8, 9 and 10, respectively.

From the above results, there can be no doubt that the product is glycerol. It is of interest to note that the glycerol in the neutral fraction did not quite match the position of authentic material if the neutral fraction from the ion-exchange columns was first run in EAW rather than BAW. However, isolation of the product from papers first run in BAW eliminated the problem (see Fig. 2C).

Glucose

The experimentally derived material (Figs. 1,3,4) matched glucose in three solvents (BAW, BPW, IBW).

Trehalose

The product matched authentic trehalose in three solvent systems (BAW, IBW, BPW) and the hydrolysis product matched glucose in BAW.

Results with C. briggsae

From the results in Table 1 and Figs. 1–4, it is clear that substantial amounts of radioactive glycerol were synthesized by *C. briggsae* only where the incubation was carried out with acetate-2-^{14}C in "whole medium". In water, little or no glycerol was detected, though good utilization of isotope occurred under both sets of conditions. From the worm extract of the "whole medium" experiment (Fig. 2A), free glycerol constituted approximately 17 per cent of the total radioactivity.

In experiments 1 and 2 (Table 1), scans of chromatograms (BAW and EAW) of the total acid products (removed from Dowex-1-acetate with HCl) showed no gross differences between the "water" and "whole medium" experiments.

Many experiments, in addition to those given in Table 1, were performed. Often, soy-peptone medium, i.e. "whole medium" but lacking liver extract, was used, with results similar to those reported. In soy-peptone medium, labeled glucose was sometimes more apparent than in Fig. 1, but it was always a minor constituent; glycerol remained the major radioactive product. "C" (Fig. 1B) was usually only visible at heavy concentrations of sample on the paper. Incubations of *C. briggsae* with acetate-2-^{14}C in buffer (0·03 M, 0·07 M or 0·1 M) or the various

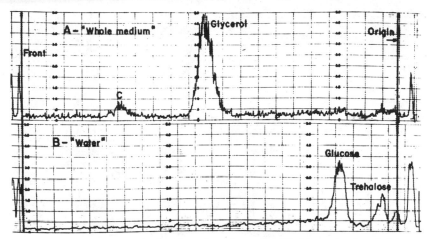

FIG. 1. Scans of the neutral products from the media of *C. briggsae* incubated with acetate-2-¹⁴C in "whole medium" and water. Chromatograms A and B were run concomitantly. Peaks before the origin and beyond the solvent front are from radioactive ink spots used for location. Solvent, BAW.

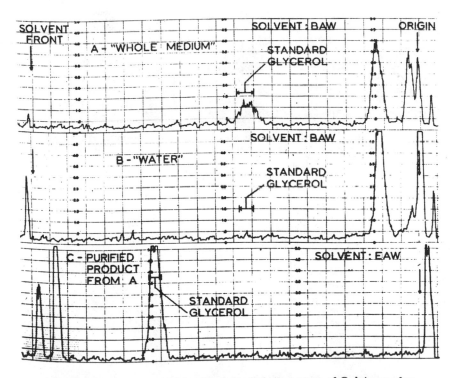

FIG. 2. Scans of the neutral fraction from alcoholic extracts of *C. briggsae* after incubation with acetate-2-¹⁴C in "whole medium" and water. Chromatograms A and B were run concomitantly. The glycerol in chromatogram C was eluted from chromatogram A. Peaks before the origin and beyond the solvent front are from radioactive ink spots used for location.

41

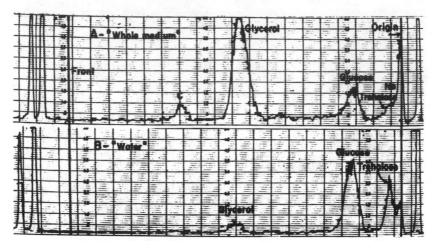

FIG. 3. Scans of the neutral products derived from the media after incubation of *T. aceti* with acetate-2-^{14}C in "whole medium" and in water. Solvent, BAW.

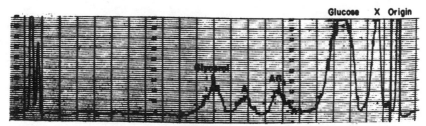

FIG. 4. Scan of the neutral products derived from "whole medium" after incubation of *P. redivivus* with acetate-2-^{14}C. "X" is not trehalose. Solvent, BAW.

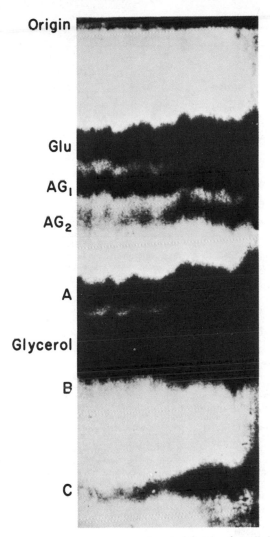

Origin

Glu

AG_1

AG_2

A

Glycerol

B

C

FIG. 5. Radioautograph of chromatogram of neutral fraction from "whole medium" after incubation of *C. briggsae* with glycerol-2-¹⁴C. "B" does not show clearly in this reproduction but can be distinguished from glycerol on the X-ray film. A control experiment in which glycerol-2-¹⁴C was incubated in the medium without nematodes showed only radioactive glycerol. Solvent, BAW.

Experiment*	Conditions†	Glycerol‡	Counts/min × 10⁻³			
			Neutral peak‡	Glu§	Asp§	Acids§
1 M	Whole medium	Yes	66			
1 WX	3·5 days	Yes	320	14	7·5	35
2 M	Water					
	3·5 days	No	23	13	6·3	
2 WX		No	410	30	6·0	58
3 M∥	Whole medium	Yes	1500¶	900	260	200
3 WH∥¶	4 days	Yes	220	3100	2000	1500
4 M∥	Whole medium	Yes	340	150	260	
4 H∥¶	6 days	Yes	208¶			
5 M∥	Water					
	3 days	No	220	110	38	330
5 WX∥		No	810	29	20	

* M = medium; WX = ethanolic extract of worms; WH = worm hydrolysate.

† The amount of acetate used in experiments 1 and 2 is given under Materials and Methods. The other results are from experiments originally performed for other purposes. In these cases, the isotope concentration typically amounted to 10–20 μc/flask and 5–20 μmoles.

‡ Where glycerol was present, it constituted the vast majority of counts in the neutral fraction from the medium (Fig. 1A); in the worm extracts, the activity of the glycerol was relatively much less, being approximately 17 per cent of the activity in the neutral peak (55,000 counts/min) in experiment 1 WX. The remaining counts were mostly in glucose and in the area near the origin of the chromatogram (Fig. 2A).

§ These figures are approximations based on aliquots of column eluates and are included only to indicate that active synthesis occurs in both water and "whole medium", and also to show the relative amount of products.

∥ Rothstein (1965).

¶ The hydrolysis procedure normally used in these experiments (6 N HCl, for 24 hr at 105°C) would cause breakdown of the glucose (and trehalose) to levulinic acid. Counts due to these compounds would therefore no longer appear in the neutral fraction. Levulinic acid elutes from the Dowex-1-acetate columns (developed with 0·5 N acetic acid), just before aspartic acid. In fact, the unidentified peak (peak A) reported to occur in hydrolysates of *C. briggsae* (Rothstein & Tomlinson, 1962) is levulinic acid. Proof of identity has been obtained by showing an exact match of the infra-red spectrum of the 2,4-dinitrophenylhydrazone of the experimentally derived material with that of authentic levulinic acid.

components of the medium (see Materials and Methods) led to little or no glycerol production either in the worms or media.

Results with T. aceti

From Fig. 3 it can be seen that the results of experiments with *T. aceti* parallel those with *C. briggsae.* The chief differences are that in "whole medium", more glucose relative to glycerol is formed, and in water radioactive glycerol is detectable. The material near the origin in Fig. 3A (whole medium) does not contain trehalose.

Results with P. redivivus

Results from *P. redivivus*, incubated with acetate-2-^{14}C in water, matched closely those obtained with *C. briggsae* and *T. aceti*. Glucose and trehalose were the chief products though small amounts of glycerol were sometimes detected. In "whole medium", however, inconsistent results were obtained. These varied from duplication of those obtained with *C. briggsae* (Fig. 1A) to those shown in Fig. 4. The products shown in Fig. 4 appear to be related to glycerol metabolism, as can be seen by comparison with Fig. 5. Peak X is not trehalose.

Incubation of C. briggsae *with glycerol-2-*^{14}C

Figure 5 shows a typical chromatogram of the neutral fraction derived from the medium in which *C. briggsae* was incubated with glycerol-2-^{14}C. Faintly radioactive areas matching the positions of AG$_1$, A, B and C have been noted from incubations of *T. aceti* with acetate-2-^{14}C ("whole medium") where large samples were chromatographed and left on X-ray film for long periods. The products formed by *P. redivivus* from acetate-2-^{14}C in "whole medium" (Fig. 4) also appear to match AG$_2$, A and C. In fact, "A's" in Figs. 4 and 5 have been shown to match in a second solvent system (IBW).

The relative amounts of the products formed from glycerol-2-^{14}C vary with the composition of the medium, though not so drastically as with acetate-2-^{14}C. For example, deletion of glucose from the medium results in a large increase in "B". AG$_1$ appears to be a sugar as determined by color sprays, e.g. diphenylamine-aniline (Smith, 1960); C is probably a glycerol ester, as hydrolysis yields glycerol; A has been tentatively identified as a 4-carbon sugar, perhaps erythrose.

DISCUSSION

It is clear that in "whole medium" (with or without added liver extract), glycerol is a major product of acetate metabolism in *C. briggsae*. In water, glycerol production is drastically reduced and glucose is the chief product. These results do not come about simply from enhanced excretion of glycerol in "whole medium", but arise from a real reduction in synthesis; little glycerol was observed inside the nematodes incubated with acetate-2-^{14}C in water (Fig. 2B). The results are also not due to simple osmolarity, since use of buffers of various concentrations led to the same results as experiments carried out in water.

Obviously, the factor causing conversion of acetate carbon into glycerol resides in one or more components of the medium. The particular component or mixture of components responsible is not known since partially constituted media (salts, vitamins, structural compounds, soy-peptone, respectively) did not result in glycerol production.

The most logical pathway for glycerol synthesis would be one consistent with recent concepts of gluconeogenesis: conversion of acetate to oxalacetate (tricarboxylic acid cycle) and thence to phosphoenolpyruvate by the action of phosphoenolpyruvate carboxykinase. The experimental results (Table 2) are in agreement with this postulate. At the stage of 3-phosphoglyceraldehyde, a branch point exists at which either glycerol (by reduction) or glucose could be formed. In "whole medium", the stress would be on synthesis of the former. In water, glycerol synthesis would be suppressed, forcing the reactions toward glucose production. Alternatively, glucose synthesis could be viewed as being suppressed in "whole medium" (which contains substantial amounts of glucose), emphasizing glycerol production as a consequence. However, one must ask why the glucose in the medium is not the source of glycerol, rather than the energetically unfavorable acetate.

TABLE 2—DISTRIBUTION OF ^{14}C IN GLYCEROL PRODUCED BY *C. briggsae*

Substrate	Glycerol (counts/min per μmole)	Carbons 1 + 3 * (counts/min per μmole)	Percentage of total	Theory†
Acetate-1-^{14}C	17,300	16,200	94	100
Acetate-1-^{14}C	17,300	17,000	98	100
Acetate-2-^{14}C	12,500	7,100	57	50
Acetate-2-^{14}C	16,000	8,150	51	50
Authentic glycerol-1,3-^{14}C	12,800	12,900	101	100

* Isolated as the dimedone derivative of formaldehyde. Values have been multiplied by 2 to correct for the fact that glycerol yields 2 molecules of formaldehyde.
† By way of the tricarboxylic acid cycle: Acetate → oxalacetate → phosphoenolpyruvate → glycerol.

The effect of malate synthetase, known to be present in the nematodes (Rothstein & Mayoh, 1966), would be to contribute extra radioactivity to carbons 1 and 3 of glycerol by way of the sequence: acetate-2-^{14}C → malate-3-^{14}C → phosphoenolpyruvate-3-^{14}C → glycerol-1,3-^{14}C. However, malate can be expected to equilibrate with succinate, thus effectively yielding malate-2,3-^{14}C, the same result as would be obtained via the tricarboxylic acid cycle. The small excess of ^{14}C found in carbons 1, 3 over the theoretical value (Table 2) may reflect the action of malate synthetase. It could also reflect fixation ^{14}CO$_2$ from the oxidation of the labeled acetate.

The remarkably similar results obtained with all three organisms studied suggest that the mechanism for turning glycerol and glucose synthesis on or off is common to free-living nematodes. A variation exists in *P. redivivus*, which in "whole medium" can produce glucose and a number of products apparently related to glycerol metabolism (see Figs. 4 and 5). Whether glycerol is a direct precursor of all or some of these compounds or whether they arise from the pathway: glycerol → glucose → products has not yet been determined. The identity of the compounds and the mechanisms of synthesis are presently being investigated.

The function of glycerol formation by free-living nematodes is obscure. Free glycerol appears in the hemolymph of certain insects at diapause in astonishing amounts (Wyatt & Meyer, 1959). Its presence has been suggested as a protective mechanism against freezing. During diapause in silkworm eggs, glycogen is converted to sorbitol and glycerol (Chino, 1958). This production may be related to the reduction in cytochrome activity occurring at this time. In *Artemia* eggs, glycerol production is related to osmotic pressure and hatching (Clegg, 1964). The above factors appear to be unrelated to free-living nematodes; most likely, glycerol production in these organisms reflects the existence of a normal and substantial gluconeogenesis pathway.

The present results re-emphasize the point made in previous reports dealing with free-living nematodes: the composition of the medium determines the nature and extent of the metabolic products formed. All research on the metabolism of these and possibily of other invertebrates in liquid media must take this factor into account.

Acknowledgements—This investigation was supported in part by Grant AI-07145 from the National Institute of Allergy and Infectious Diseases, United States Public Health Service.

ABRAHAM S. & HASSID W. Z. (1957) *Methods in Enzymology* (Edited by COLOWICK S. P. & KAPLAN N. O.), Vol. IV, p. 527. Academic Press, New York.

CHINO H. (1958) Carbohydrate metabolism in diapause eggs of the silkworm, *Bombyx mori*— II. Conversion of glycogen to sorbitol and glycerol during diapause. *J. Insect Physiol.* 2, 1–12.

CLEGG J. S. (1964) The control of emergence and metabolism by external osmotic pressure and the role of free glycerol in developing cysts of *Artemia salina*. *J. exp. Biol.* 41, 879–892.

HIRS C. H., MOORE S. & STEIN W. H. (1954) The chromatography of amino acids on ion exchange resins. *J. Am. chem. Soc.* 76, 6063–6065.

ROTHSTEIN M. (1963) Nematode biochemistry—III. Excretion products. *Comp. Biochem. Physiol.* 9, 51–59.

ROTHSTEIN M. (1965) Nematode biochemistry—V. Intermediary metabolism and amino acid interconversions in *Caenorhabditis briggsae*. *Comp. Biochem. Physiol.* 14, 541–552.

ROTHSTEIN M. & COOK E. (1966) Nematode biochemistry—VI. Conditions for axenic culture of *Turbatrix aceti*, *Panagrellus redivivus*, *Rhabditis anomala* and *Caenorhabditis briggsae*. *Comp. Biochem. Physiol.* 17, 683–692.

ROTHSTEIN M. & MAYOH H. (1964) Nematode biochemistry—IV. On isocitrate lyase in *Caenorhabditis briggsae*. *Archs Biochem. Biophys.* 108, 134–142.

ROTHSTEIN M. & MAYOH H. (1966) Nematode biochemistry—VIII. Malate synthetase. *Comp. Biochem. Physiol.* 17, 1181–1188.

ROTHSTEIN M. & TOMLINSON G. (1962) Nematode biochemistry—II. Biosynthesis of amino acids. *Biochim. biophys. Acta* 63, 471–480.

SAYRE F. W., HANSEN E. L. & YARWOOD E. A. (1963) Biochemical aspects of the nutrition of *Caenorhabditis briggsae*. *Expl Parasit.* 13, 98–107.

SMITH I. (1960) *Chromatographic and Electrophoretic Techniques*, p. 251. Interscience, New York.

TOMLINSON G. A. & ROTHSTEIN M. (1962) Nematode biochemistry—I. Culture methods. *Biochim. biophys. Acta* 63, 465–470.

WYATT G. R. & MEYER W. L. (1959) The chemistry of insect hemolymph. *J. gen. Physiol.* 42, 1005–1011.

Key Word Index—Nematode biochemistry; *Caenorhabditis briggsae*; *Turbatrix aceti*; *Panagrellus redivivus*; glycerol; acetate-14C; metabolism.

FUNCTIONAL MORPHOLOGY AND HISTOCHEMISTRY OF STRUCTURAL PROTEINS OF THE GENITAL CONE OF *COOPERIA PUNCTATA* (VON LINSTOW, 1907) RANSOM, 1907, A NEMATODE PARASITE OF RUMINANTS

Frank Stringfellow

The genital cone, a complex structure at the posterior end of many male nematodes, has been described in some species and used to a limited extent as a generic and specific taxonomic character. However, the extent to which the genital cone can be fully utilized in taxonomic investigations is uncertain because its structure and variations are little known. The genital cone is intimately associated with the spicules, the conformation of which is used as a prime character in trichostrongylid taxonomy. Moreover, differences in the spicules of the species of *Cooperia* have been established but little is known about their genital cones. The common intestinal species of cattle and sheep, *C. punctata* (von Linstow, 1907) Ransom, 1907, was chosen for this investigation.

Existing illustrations of the genital cone are incomplete (Baylis, 1929); or, if the gross features of the genital cone are illustrated adequately, details, which are important for distinguishing between species, are not included (Cram, 1925). At present, the best description (Andreeva, 1958) of this structure in this genus is based on *Cooperia oncophora*. The present work describes the morphology of the genital cone in *C. punctata*, its function as observed in live specimens, and its partial chemical characterization as determined histochemically.

MATERIALS AND METHODS

Functional morphology

The genital cones of 156 *C. punctata* of bovine origin were studied. Some had been stored in a preservative of 92 parts 70% ethanol, 3 parts formalin, and 5 parts glycerine. This material was prepared for study from the caudal, lateral, dorsal, and ventral aspects, by using glycerine, lactophenol, and rosaniline-hydrochloride, respectively. The procedure for staining with rosaniline-hydrochloride was as follows: *C. punctata* were heated in a saturated solution of rosaniline-hydrochloride in lacto-phenol. The worms were heated on a hot plate until vapors rose from the lacto-phenol–stain mixture and then were allowed to cool for 5 min. This procedure was repeated 3 times. Excess stain was rinsed from the worms with 70% ethanol. They were transferred to lacto-phenol until destained to an optimal density. Surface features of the genital cone were studied in whole mounts with the AgNO₃ precipitation technique for studying epidermal plates of miracidia (Cable, 1961). All measurements in Table I were determined with the aid of a calibrated ocular micrometer from both whole mounts and sectioned worms.

Specimens of *C. punctata* were collected from the small intestine of a year-old, untreated steer and were studied live in isotonic saline without staining. Retraction and extrusion of the spicules and associated changes occurring in the genital cone were studied. No worms in copula were observed.

Histochemistry

Procedures used in this study are as given in the Manual of the Armed Forces Institute of Pathology (AFIP) (1960), Barka and Anderson (1965), and Pearse (1960) unless stated otherwise. Paraffin-embedded sections of worms fixed in 10% formalin were stained with the periodic acid-Schiff method. Acetylation, unblocking by saponification, and exposing sections to Schiff's reagent without prior periodic acid oxidation were used as controls to demonstrate the specificity of the reaction. Sections were exposed to digestion with 1.0% malt diastase in 0.85% NaCl for 45 min at 37 C. Metachromasia was detected with toluidine blue before and after digestion with testicular hyaluronidase (General Biochemicals (GB): 1 mg/ml in 0.85% NaCl at 37 C for 1 hr). The carbohydrate moiety was further characterized with alcian blue (Allied Chemicals: C.I. 74240: lot: 0465147).

The pyronin-methyl green method was used to stain for nucleic acids in worms fixed in Carnoy's fluid. Controls consisted of sections digested with ribonuclease before staining (GB: at 1 mg/ml in distilled H_2O for 1 hr at 37 C). The methyl green dye was extracted by shaking its aqueous solution with an excess of chloroform and allowing it to stand for 3 days.

Sudan black B in ethanol was used as a general lipid stain and bound lipids were detected using Berenbaum's method. The worms were fixed in formalin, embedded in polyethylene glycol, and sectioned. Controls were extracted with chloroform:methanol (1:3) as suggested by Barka and Anderson (1965).

Proteins were detected with the mercury bromphenol blue method. Proteins containing tyrosine were detected using Bensley and Gersh's modification of Millon's reaction, tryptophane with Adam's DMAB-nitrite method, and arginine with the Sakaguchi reaction. Collagen was detected with Mallory's aniline blue collagen stain and Van Gieson's staining with and without digestion with collagenase (GB: 1 mg/ml in 0.85% NaCl at 50 C for 3 hr). Weigert's and Verhoeff's procedures were each used as a test for elastin. Gridley's reticulum stain and Foot's modification of Bielchowsky's technique were used to test for reticulin. The dihydroxy dinaphthyl disulfide method was used to detect SH groups. Disulfide bonds were detected by reduction with sodium thioglycolate and subsequently staining for SH groups. The performic acid alcian blue test was used to detect keratin (Pearse, 1960). The genital cone was incubated in pepsin (0.2 mg/ml in 0.02 N HCl, pH 2.4 at 37 C for 20 hr), trypsin (0.2 mg/ml in 0.05 M phosphate buffer, pH 8.2 at 37 C for 20 hr), and papain (10 mg/ml in 0.05 M phosphate buffer at pH 8.2 at 37 C for 20 hr) to determine its resistance to digestion with proteolytic enzymes. Solubility of the genital cone was tested in 0.1 N HCl and 0.1 N KOH solutions at 37 C for 20 hr. Fontana's silver staining for phenolic compounds, incubation of tissue sections in 3,4 dioxyphenylalanine to detect DOPA oxidase, and Johri and Smyth's method (1956) for phenolic compounds were used to detect quinone tanning. All worms were fixed in 10% formalin except those subjected to Mallory's aniline blue collagen stain which were fixed in Zenker's fluid, and those enzymatically digested which were fixed in Carnoy's fluid.

Mammalian tissue controls were processed and studied concurrently with the worms. All tissues were sectioned at 3 μ and only dyes certified by the Biological Stain Commission were used in this study.

RESULTS

Functional morphology

Rosaniline-hydrochloride staining, which is not critical, greatly facilitates study of the genital cone. The dye stains the nematode generally much as cotton blue (methyl blue) and does not stain the genital cone specifically. The cuticle of the worm is dulled so that the glaring effect produced by light reflected from the cuticle of a lacto-phenol-cleared nematode is considerably reduced. Rosaniline-hydrochloride was superior to other triphenyl methane derivatives that were tried (malachite green, light green, fast green, acid fuchsin, methyl blue, crystal violet, and aniline blue). Using this new staining technique enables observation of about 90% of the genital cone in whole mounts; however, study of sections of the nematode stained with Mallory's aniline blue collagen stain should be made to confirm structure. Acid fuchsin selectively stains the sclerotized parts red, whereas the aniline blue–orange G solution stains the cuticular areas blue (Fig. 7). Hematoxylin (Heidenhain's) and eosin (Fig. 10) and Masson's trichrome stain may also be used. Hematein stains the sclerotized parts black whereas the cuticular areas stain red with eosin. In Masson's trichrome stain, the Biebrich scarlet–acid fuchsin solution intensely stains sclerotized parts and light green stains cuticular areas.

Description of genital cone (Figs. 1–6; Table I)

Reduced *accessory bursal membrane* (I, K, X) at posterior end of body bordered on anterior surface by sclerotization. Dorso-laterally, accessory bursal membrane joined by paired *dorsal raylets*

→

FIGURES 1–6. Genital cone of *Cooperia punctata*. 1. Ventral. 2. Lateral. 3. End-on view of posterior end. 4. Dorsal view of genital cone. 5. Ventral view of accessory bursal membrane. 6. Stereogram of genital cone. Bursa and spicules are not shown. All drawings except Figure 6 made with the aid of a camera lucida. Refer to Table I for anatomical terms corresponding to letters.

49

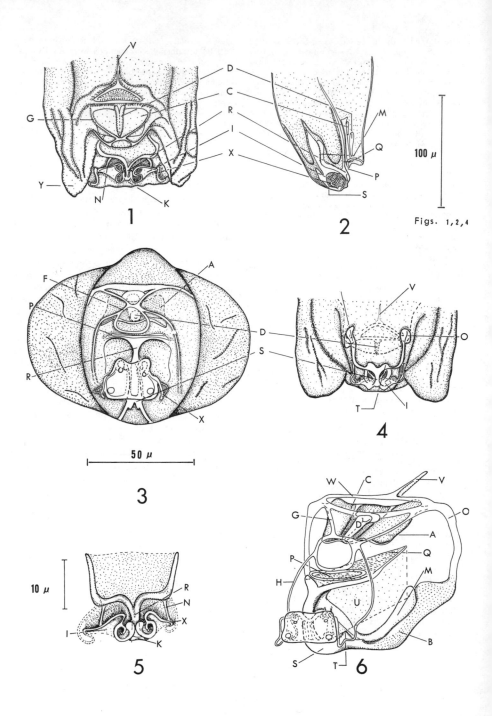

TABLE I. *Comparison of terminology of Andreeva and the author for genital cones of* C. oncophora *and* C. punctata.

Lr.	Andreeva's terminology	Author's terminology	Range* (μ)	
			L	W
A	ventro-ventral plate of the basal apparatus	dorso-ventral plate	26.2–33.3	17.8–24.1
B	dorso-dorsal plate	dorso-dorsal plate	11.0–25.2	
C	ventral plate	ventral plate	18.5–23.8	
D	accretion of ventral raylet of genital cone	ventral raylet	12.1–18.2	4.0–4.0
E	lateral cuticular protuberance of genital cone	not present		
F	semicircular band of distal formation	circular cuticular ridge		
G	not present	ventral cuticular ridge		
H	not present	postero-lateral cuticular ridge		
I	horizontal lying plate of distal formation	horizontal plate		8.0–16.2
J	dorsal papilla	not present		
K	distal region of distal formation	accessory bursal membrane		
L	spicule	spicule		
M		ventro-dorsal plate		
N		dorsal raylets		
O		baso-dorsal plate connection		
P		cloacal opening	17.2–21.2	
Q		cloacal plate	28.3–34.3	
R		ventral sclerotized support		
S		dorsal sclerotized support		
T		dorsal flange of dorsal plate		
U		septum		
V		anterior projection of cuticular ridge		
W		cuticular ridge		
X		sclerotized bars		
Y		lateral cuticular masses		

* Measurements are maximum values.

(N). Central sclerotized piece medially connecting dorsal and ventral sclerotized borders. *Horizontal plates* (I) two-thirds ventrally on accessory bursal membrane (I, K, X). Paired sclerotized flanges located medially on anterior surface of horizontal plate. Each flange curves posterodorsally and bifurcates into a short branch which enters posterior wall of cup of accessory bursal membrane (I, K, X), and a long branch which curves antero-ventrally then laterally flexing dorsally over lateral border of accessory bursal membrane (I, K, X) as a *sclerotized bar* (X). This bar (X) extends dorsally beyond dorsal border of accessory bursal membrane (I, K, X) where it curves ventro-medially and then, flush with the cuticle, extends dorso-laterally after which it (H) extends ventrally joining the *circular cuticular ridge* (F). Remainder of genital cone flush with cuticle of the nematode or within its body. Anterior surface of accessory bursal membrane (I, K, X) joins the *ventral sclerotized support* (R). Ventral sclerotized support (R) projects anteriorly bifurcating into sclerotized thickenings extending antero-dorsally, dorsal to, and underlying cloaca as a plate. This *cloacal plate* (Q) is arrow-shaped, broad at its posterior extremity, and gradually narrows anteriorly. Dorso-medially, *septum* (U) extends from cloacal plate (Q) to cuticle. Ventral to ventral sclerotized support is a horizontal slit, the *cloacal opening* (P). Cloacal opening (P) dorsal to circular cuticular ridge (F). An anterior projection (V) extends from *cuticular ridge* (W).

This ridge connects to circular cuticular ridge (F) by a cuticular extension (G). *Ventral raylet* (D) extends from posterior border of cuticular ridge (W) and lies in groove of basal apparatus (A, C, D).

Basal apparatus (A, C, D) consists of *ventral* (C), *dorso-ventral* (A) *plates*, and ventral raylet (D). Dorso-ventral plate (A) underlies and extends from circular cuticular ridge (F) to slightly anterior to cuticular ridge (W) where it extends dorsally as two antero-lateral processes. Posteriorly, dorso-ventral plate (A) is shaped like a flattened M forming a medial groove in which ventral raylet (D) lies. Dorso-ventral plate (A) at antero-lateral corner forms a connection with dorso-dorsal plate (B), the *baso-dorsal plate connection* (O). Ventral plate (C) originates from anterior one-third of dorso-ventral plate (A) and with dorso-ventral plate forms groove in which ventral raylet (D) lies. Ventral raylet (D) has knoblike base attaching to posterior border of cuticular ridge (W). A conical projection fits onto this base and contains a central core that attaches to the knoblike base.

Dorsal sclerotized support (S) extends antero-dorsally from accessory bursal membrane (I, K, X). It bifurcates, each branch bending laterally then anteriorly, forming dorso-dorsal (B) plate. A flange extending dorsally is medial to the first bend. Ventro-medial to the dorso-dorsal plate a flange, the *ventro-dorsal plate* (M), extends posteriorly to the dorsal sclerotized support (S). Ex-

51

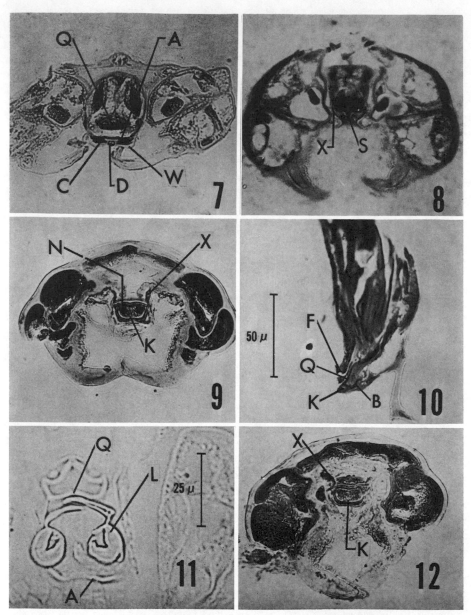

FIGURES 7–12. Stained sections of *Cooperia punctata*. **7.** Mallory's aniline blue collagen stain—cross section. **8.** Bromphenol blue for proteins—tangential section. **9.** DDD method for SS bonds (after reduction)—cross section. **10.** H and E—longitudinal section. **11.** Millon's reaction for tyrosine—tangential—cross section. **12.** DDD method for SH groups—cross section.

tending from ventro-dorsal plate (M) to accessory bursal membrane (I, K, X) is the dorsal raylet (N). A *lateral cuticular mass* (Y) forms each side of this genital cone.

Figure 7 shows muscles attached to the cloacal plate (Q). On each side of the septum are muscles that insert at the junction of the ventral (R) and dorsal (S) sclerotized supports.

The dorso-dorsal plate (B) and attached structures are pulled anteriorly during extrusion of the spicules. The anterior portion of the cloacal plate (Q) guides the distal portion of the spicules ventrally so that the distal portion of the spicules is forced against the anterior region of the dorso-ventral plate (A). At this time, the genital cone is somewhat flexed ventrally. The spicules slide posteriorly using the dorso-ventral plate (A), cloacal plate (Q), and the cloacal lips (P) as guides to the exterior. When partially out, the dorsal flange of the spicules forces the shaft against the ventral portion of the genital cone so that the ventral raylet (D) appears erect. The left spicule usually is extruded first, followed by the right spicule.

Figures 7 and 11 show that the genital cone serves as a skeletonlike support for the posterior end as well as for the cloacal walls.

Histochemistry

The results of all histochemical tests are summarized in Tables II and III.

DISCUSSION

Functional morphology

There was no significant variation in the structure of the genital cone in the specimens of *C. punctata* examined. This fact indicates that the genital cone is a stable character that may be used as another taxonomic character not only within the genus *Cooperia* but possibly within other genera. Its use as a taxonomic character in this genus has been analyzed (Stringfellow, in preparation) similar to the bosses in *Nematodirus* from sheep (Stringfellow, 1968). Subsequent to establishing genital cone morphology is the need for designing brief, accurate terminology. Some additions and changes have been made herein for accuracy and clarity in Andreeva's (1958) proposed terminology, which is also given in Table I. For example, it is more accurate to describe the ventro-ventral plate of the basal apparatus as the dorso-ventral plate. Also, Andreeva (1958) indicates the basal apparatus is a separate entity from the dorsal portion of the genital cone; however, histological studies show that the basal apparatus in some *Cooperia* is associated with the rest of the genital cone via the baso-dorsal plate connection. More terminology may be introduced to describe features of the genital cone present in species other than *C. punctata*. The genital cone of *C. punctata* should not be considered general enough to possess all of the structures found in this genus. For example, Andreeva (1958) illustrates the lateral cuticular protuberance (Table I) in *C. oncophora* whereas *C. punctata* does not have this structure.

Most authors describe only the sclerotized parts of the genital cone; however, there is an interconnection between the cuticle and the sclerotized parts. I believe (Figs. 1–6) cuticular involvement in the genital cone should be included along with the sclerotized parts as was shown, not only in sections but also by the silver precipitation technique. When examining Figure 6, it must be understood that a sheet of cuticle covers most of the genital cone shown except for those sclerotized structures that are external and those cuticular areas that are raised from the rest of the cuticle. Tissue sections stained with Mallory's aniline blue collagen stain in conjunction with rosaniline-hydrochloride staining of worm whole mounts should be used in studying the genital cone. About 90% of the morphology can be determined with rosaniline-hydrochloride staining provided the specimens are suitable.

Since Hall (1921) first described the telamon in *Hyostrongylus rubidus*, the relation of the genital cone to the telamon has been confused. Cram (1925) described the sclerotized parts of the genital cone as a telamon, whereas Baylis (1920) described a similar structure in two new species of *Cooperia* and called it a genital cone. Andreeva (1958) considers the genital cone to be the cone-shaped structure containing the muscle–nerve complex and supporting elements. Andreeva considers the telamon to be the supporting apparatus lying in the ventral part of the genital cone and skeletonizing the ventral wall of the cloaca but at times found dorsally to the cloaca. I basically agree with the latter individual but emphasize that these designations (telamon, genital cone) are simply a matter of degree of sclerotization. The telamon and the genital cone are interrelated and continuous with one another.

The genital cone serves three functions: (1) site for muscle attachment; (2) guide for ex-

TABLE II. *Histochemistry of genital cone of* C. punctata *carbohydrates, nucleic acids, and lipids.*

Histochemical test	A	B	C	D	E	F	G	H	I	J	K	L	M	N	O	P	Q	R	S	T	U	V	W	X	Y	Reference
Carbohydrates																										
Periodic acid-Schiff	+	−	−	−	+	+	+	−	−	−	−	−	+	−	−	−	+	+	+	−	+					AFIP*
acetylation	−	−	−	−	−	−	−	−	−	−	−	−	−	−	−	−	−	−	−	−						B & A**
saponification after acetylation	−	−	−	−	+	+	+	−	−	−	−	−	−	−	−	−	+	+	−	+						B & A
no periodic acid oxidation	−	−	−	−	−	−	−	−	−	−	−	−	−	−	−	−	−	−	−	−						B & A
diastase	−	−	−	−	+	+	+	−	−	−	−	−	−	−	−	−	+	−	+	−	+					AFIP
Metachromasia	−	−	−	−	−	−	−	+	+	−	+	−														AFIP
hyaluronidase digestion	−	−	−	−	−	−	−	+	+	−	+	−														AFIP
Alcian blue	−	−	−	−	−	−	−	−	−	−	−	−	−	−	−	−	−	−	−	−						AFIP
hyaluronidase digestion	−	−	−	−	−	−	−	−	−	−	−	−	−	−	−	−	−	−	−	−						AFIP
Nucleic Acids																										
Methyl green-Pyronin																										
RNA	−	−	−	−	+	+	+	−	−	−	−	−	−	−	−	−	−	−	+	−	−					B & A
DNA	−	−	−	−	−	−	−	−	−	−	−	−	−	−	−	−	−	−	−	−						B & A
RNase	−	−	−	−	−	−	−	−	−	−	−	−	−	−	−	−	−	−	−	−						B & A
Hematoxylin and eosin	+	+	+	+	+	+	+	+	+	+	+	+	+	+	+	+	+	+	+	+	+	+				AFIP
Lipids																										
Sudan black	−	−	−	−	−	−	−	−	−	−	−	−	−	−	−	−	−	−	−	−						B & A
Berenbaum (bound lipid)	−	−	−	−	−	−	−	−	−	−	−	−	−	−	−	−	−	−	−	−						B & A

* Manual of the Armed Forces Institute of Pathology (1960).
** Barka and Anderson (1965).
  Symbols: + = faint staining; − = no staining.

truding and retracting the spicules; and (3) a skeletonlike support for the posterior end as well as for the cloacal walls. The genital cone is a movable structure that is flexed by contraction of the muscles inserted on it. During spicule extrusion and retraction, flexure of the genital cone permits movement of the spicules in and out of the cloaca in this species of *Cooperia.* The ventral, dorso-ventral plates, as well as the cloacal plate serve as a funnel

TABLE III. *Histochemistry of genital cone of* C. punctata *proteins.*

Histochemical test	A	B	C	D	E	F	G	H	I	J	K	L	M	N	O	P	Q	R	S	T	U	V	W	X	Y	Reference
Mercury bromphenol blue	×	×	×	×	+	+	+	+	×	×	×	×	×	×	×	×	+	+	+	×	×					B & A*
DMAB nitrite	+	+	+	+	−	−	−	+	+	+	+	+	+	+	+	+	+	+	−	−	+	+				B & A
Sakaguchi	−	−	−	−	+	+	+	−	−	−	−	−	−	−	−	−	−	−	−	−						B & A
Millon's test	+	+	+	+	−	−	−	+	+	+	+	+	+	+	+	+	+	+	−	−	−	−				B & A
Mallory's aniline blue	×	×	×	×	+	+	+	×	×	×	×	×	×	×	×	×	+	+	+	×	+					AFIP**
Van Gieson's	+	+	+	+	+	+	+	+	+	+	+	+	+	+	+	+	+	+	+	+	+	+				AFIP
Collagenase digestion	+	+	+	+	−	−	−	+	+	+	+	+	+	+	+	+	+	+	+	+	+					
Weigert's	−	−	−	−	−	−	−	−	−	−	−	−	−	−	−	−	−	−	−	−						AFIP
Verhoeff's	−	−	−	−	−	−	−	−	−	−	−	−	−	−	−	−	−	−	−	−						AFIP
Gridley	−	−	−	−	−	−	−	−	−	−	−	−	−	−	−	−	−	−	−	−						AFIP
Foot's modification of Bielchowsky	−	−	−	−	−	−	−	−	−	−	−	−	−	−	−	−	−	−	−	−						AFIP
Fontana's silver	−	−	−	−	−	−	−	−	−	−	−	−	−	−	−	−	−	−	−	−						AFIP
DOPA reaction	−	−	−	−	−	−	−	−	−	−	−	−	−	−	−	−	−	−	−	−						AFIP
Johri and Smyth (1956)	−	−	−	−	−	−	−	−	−	−	−	−	−	−	−	−	−	−	−	−						
Masson's trichrome	×	×	×	×	+	+	+	+	+	×	×	×	×	×	×	×	+	+	+	+	+					AFIP
PFAAB	+	+	+	+	+	+	+	+	+	+	+	+	+	+	+	+	+	+	+	+	+	+				P†
SH	×	×	×	×	−	−	−	×	×	×	×	×	×	×	×	×	×	−	−	×	×					B & A
SS	×	×	×	×	×	×	×	×	×	×	×	×	×	×	×	×	×	×	×	×	×	×				B & A
Papain	−	−	−	−	−	−	−	−	−	−	−	−	−	−	−	−	−	−	−	−	−					K & D††
Trypsin	−	−	−	−	−	−	−	−	−	−	−	−	−	−	−	−	−	−	−	−						K & D
Pepsin	−	−	−	−	−	−	−	−	−	−	−	−	−	−	−	−	−	−	−	−						K & D
KOH	−	−	−	−	−	−	−	−	−	−	−	−	−	−	−	−	−	−	−	−						
HCl	−	−	−	−	−	−	−	−	−	−	−	−	−	−	−	−	−	−	−	−						

* Barka and Anderson (1965).
** Manual of the Armed Forces Institute of Pathology (1960).
† Pearse (1960).
†† Kan and Davey (1968).
  Symbols: × = intense staining; + = faint staining; − = no staining.

for mechanically channeling the spicules to the exterior (Fig. 11). This involves support and protection of the cloacal floor and roof so that the spicules will not be pushed through them.

Histochemistry

Cuticular areas stained lightly by the PAS reagent indicate either a low content of PAS-positive substances or material not giving a strong positive reaction (mucopolysaccharides and glycoproteins). The absence of PAS staining after acetylation for 90 min indicates a lack of unsaturated lipids. Generally, the PAS reaction of unsaturated lipids cannot be prevented even by 24 hr acetylation. The genital cone did not stain with alcian blue, indicating that PAS staining of cuticular areas was not caused by acid mucopolysaccharides. Beta metachromasia may indicate more the presence of negatively charged radicals than acid mucopolysaccharides. Little carbohydrate material was detected in the genital cone and probably a collagenlike protein caused the positive PAS reaction of the cuticle. Collagen contains appreciable amounts of polysaccharide material which may be combined with the protein moiety. The exact composition of the polysaccharide moiety is not known but hexoses and hexosamines are present.

Nucleic acids and lipids were not detected in the sclerotized parts of the genital cone. In general, the keratins of wool, hair, horn, hoof, and feathers, when freed from extraneous material, consist entirely of protein. They do not contain RNA, DNA, ascorbic acid, alkaline phosphatase, or glutathione; therefore, it is understandable that they were not detected. Anya (1966a) detected RNA in the nematode cuticle, indicating that the cuticle is a metabolic structure with turnover rather than an inert secretion. This fact was further substantiated in this study.

Millon's reaction, which can be considered a general protein stain, stained the sclerotized parts of the genital cone but not the cuticular areas. The mercury bromphenol blue method stained protein in cuticular and sclerotized parts of the genital cone (Fig. 8). Proteins deficient in tyrosine give a negative Millon's reaction. Collagen has a low percentage of tyrosine as is indicated by the lack of staining

of cuticular portions of the genital cone with Millon's test. The cuticular parts of the genital cone did not stain with the DMAB-nitrite test for tryptophane. Collagen, in general, is deficient in tryptophane. The absence of staining using Weigert's, Verhoeff's, Gridley's, and Foot's modification of Bielchowsky's technique further supports differentiation of collagen from other fibrous proteins in cuticular parts. Cuticular areas of the genital cone stained positively for arginine using the Sakaguchi reaction whereas sclerotized areas did not. Herlich (1966) detected arginine in hydrolysates of whole C. punctata; however, tyrosine was not detected and tryptophane was not determined.

Mallory's aniline blue (Fig. 7) and Van Gieson's stains detected collagen in the cuticular parts of the genital cone. Although these stains are, at best, empirical, they are meaningful when used in conjunction with collagenase digestion of sections. Anya (1966b) detected collagen in the cuticle of Ascaris lumbricoides and Chitwood (1936) proposed the term ascarocollagen for it. In vertebrate tissues, collagen consists of fibril bundles showing an axial periodicity of 640 Å. However, Watson (1958) showed that not all chemically defined collagens show this periodicity. Moreover, Schmidt et al. (1955) indicated that axial periodicity was a reflection of an environmental influence on the protofibrils during their formation. Fibrils of the nematode cuticle generally lack this periodicity (Hinz, 1963; Watson, 1965), indicating that they are collagenlike rather than true collagen.

Quinone tanning has been detected in the nematode cuticle (Brown, 1950). Polyphenol oxidase represents a pair of enzyme activities occurring together and catalyzing the oxidation of tyrosine to DOPA quinone. DOPA quinone through a series of intermediate compounds leads to quinone tanning. These compounds were not detected in the genital cone of C. punctata; therefore, quinone tanning probably does not enter into the sclerotization process.

Invertebrate proteins containing sulfur generally are called keratins. However, there are many proteins with sulfur and there is much variation within each type (Brown, 1950; Krishnan, 1953; Trim, 1941; Monné, 1956). Scleroproteins in closely related animals may differ in amino acid composition and a single scleroprotein may have a different amino acid composition at different stages in the life history of the organism. This fact makes it difficult to define a keratin and may account for the absence of arginine in the sclerotized parts of the genital cone. During keratinization, the SH groups of cysteine are oxidized to the SS bonds and chemical stability may be used to distiguish scleroproteins from less stable sulfur-containing proteins such as insulin. Keratin as a component of sclerotized parts of the genital cone was indicated by: (1) resistance to digestion by trypsin and pepsin; (2) relative insolubility in KOH and HCl; (3) a positive PFAAB and Masson's trichrome stain; and (4) the presence of SS bonds (Fig. 9) and SH groups (Fig. 12) in the sclerotized parts of the genital cone. Relative resistance of the genital cone to digestion by proteolytic enzymes and its insolubility in dilute acid and alkali solutions is attributed to the structure of the keratin molecule. Keratin consists of closely packed polypeptide chains held together by the SS bonds of cysteine. Resistance to certain solvents is caused by the relatively high degree of cross-linking of the peptide chains by the SS bonds. Resistance to digestion by trypsin and pepsin is associated with the close packing of the chains. Masson's trichrome stain and Mallory's aniline blue collagen stain are important indicators of keratin because keratin fibers are erythrophilic. A tissue element is said to be strongly erythrophilic when it has a marked affinity for the dye acid fuchsin. Generally, the oldest fibers formed by fibroproteins are strongly erythrophilic. This fact then explains the selective staining of the sclerotized parts of the genital cone in section with acid fuchsin. Keratin satisfies the functional requirements of the genital cone because of its tensile strength attributed to the large number of similar peptide chains in parallel and its elasticity explained by folding and unfolding of the chains.

The inability of the sclerotized parts to stain for arginine can be attributed to two causes: (1) a low concentration or absence of arginine and (2) the necessity of having to use thin sections (3 μ) for studying the genital cone. A low concentration of arginine would be unusual since arginine, along with other amino acids, is a primary constituent of the keratin molecule. Since the Sakaguchi reaction is faint, it is likely that the use of thin sections can account for the inability to demonstrate arginine in the sclerotized parts of the genital cone.

ACKNOWLEDGMENTS

I wish to thank Dr. Harry Herlich, Beltsville Parasitological Laboratory, for supplying the live *Cooperia punctata* used in this study and Miss Judith Humphrey for translating terminology from Russian to English.

LITERATURE CITED

ANDREEVA, N. K. 1958. Atlas of Helminths (Strongylata) of Domestic and Wild Ruminants of Kazakhstan. Tashkent, 215 p.

ANYA, A. O. 1966a. Localization of ribonucleic acid in the cuticle of nematodes. Nature **209**: 827–828.

———. 1966b. The structure and chemical composition of the nematode cuticle. Observations on some oxyurids and *Ascaris*. Parasitology **56**: 179–198.

ARMED FORCES INSTITUTE OF PATHOLOGY. 1960. Manual of Histologic and Special Staining Techniques. McGraw-Hill, New York, 207 p.

BARKA, T., AND P. J. ANDERSON. 1965. Histochemistry Theory, Practice, and Bibliography. Harper and Row, New York, 660 p.

BAYLIS, H. A. 1929. Two new species of *Cooperia* (Nematoda) from Australian cattle. Ann. Mag. Nat. Hist., London **4**: 529–533.

BROWN, C. H. 1950. Quinone tanning in the animal kingdom. Nature **165**: 275.

CABLE, R. M. 1961. An Illustrated Laboratory Manual of Parasitology. Burgess, Minneapolis, Minnesota, 165 p.

CHITWOOD, B. G. 1936. Observations on the chemical nature of the cuticle of *Ascaris lumbricoides* var *suis*. Proc. Helm. Soc. Wash. **3**: 39–49.

CRAM, E. B. 1925. *Cooperia bisonis*, a new nematode from the buffalo. J. Agr. Res. **30**: 571–573.

HALL, M. C. 1921. Two new genera of nematodes, with a note on a neglected nematode structure. Proc. U. S. Natl. Mus. **59**: 541–546.

HERLICH, H. 1966. Amino acid composition of some strongyle parasites of cattle. Proc. Helm. Soc. Wash. **33**: 103–105.

HINZ, E. 1963. Elektronmikroskopische Untersuchungen an *Parascaris equorum*. Protoplasm **56**: 202–241.

JOHRI, L. N., AND J. D. SMYTH. 1956. A histochemical approach to the study of helminth morphology. Parasitology **46**: 107–116.

KAN, S. P., AND K. G. DAVEY. 1968. Molting in a parasitic nematode, *Phocanema decipiens*.

II. Histochemical study of the larval and adult cuticle. Can. J. Zool. **46**: 235–241.

KRISHNAN, G. 1953. On the cuticle of the scorpion *Palamneus swammerdami*. Quar. J. Micr. Sci. **94**: 11–21.

MONNÉ, L. 1956. On the histochemical properties of the egg envelopes and external cuticles of some parasitic nematodes. Ark. Zool. **9**: 93–113.

PEARSE, A. G. E. 1960. Histochemistry Theoretical and Applied. Churchill Ltd., London.

SCHMIDT, F. O., J. GROSS, AND J. H. HIGHBERGER. 1955. States of aggregation of collagen. Symp. Soc. Exp. Biol. **9**: 148–162.

STRINGFELLOW, F. 1968. Bursal bosses as a taxonomic character in six species of *Nematodirus* from domestic sheep, *Ovis aries*, in the United States. J. Parasit. **54**: 891–895.

TRIM, A. C. H. 1941. The protein of the insect cuticle. Nature, London **147**: 115–116.

WATSON, B. D. 1965. The fine structure of the body wall and the growth of the cuticle in the adult nematode *Ascaris lumbricoides*. Quar. J. Micr. Sci. **106**: 83–91.

WATSON, M. R. 1958. The chemical composition of the earthworm cuticle. Biochem. J. **68**: 416–420.

Nematode Physiology

EFFECTS OF TEMPERATURE ON THE DEVELOPMENT OF EGGS OF *NEMATOSPIROIDES DUBIUS* UNDER AXENIC CONDITIONS RELATIVE TO IN VITRO CULTIVATION

Kazuo Yasuraoka and Paul P. Weinstein

In a study of the reproductive behavior of *N. dubius* in vitro, Sommerville and Weinstein (1964) reported some females were apparently fertilized by the males, but that invariably the eggs harvested from the medium did not develop beyond early cleavage. In that study, eggs deposited at 37 C were collected once daily, and were subsequently maintained at 18 to 22 C. It appeared possible that the failure to obtain complete embryonation and hatching might have been due to the fact that the eggs thus exposed to 37 C for approximately 24 hr were injured. This paper reports observations on the effect of temperature on the development of *N. dubius* eggs under axenic conditions.

MATERIALS AND METHODS

To obtain worms as a source of eggs, male mice (NIH "general purpose" strain) were each infected with 200 to 400 filariform larvae. Mature adults were recovered from the intestines and thoroughly washed in saline containing antibiotics, following the procedures previously described (Sommerville and Weinstein, 1964), except for some modifications noted below. In those experiments in which worms or the eggs obtained from them were to be exposed to 37 C as a test condition, worms were recovered from the intestines, and cultures were prepared in a warm room maintained at 36 to 37 C. This assured that neither worms nor eggs were exposed to a drop in temperature prior to the initiation of the test condition. The uterus of such worms contained eggs in various stages of development, ranging from newly fertilized ova in the anterior portion to eggs containing embryos in the 2- to 8-cell stage in the posterior uterus.

Preparation of cultures with eggs deposited by adult worms: Female worms were recovered from mice infected for approximately 2 months, and were inoculated into the following media: 1) NCTC 109 (McQuilkin, Evans, and Earle, 1957) alone or containing 20% fetal bovine serum; 2) chick embryo extract, human serum and a vitamin mixture (Weinstein and Jones, 1959). All media contained 200 units penicillin and 100 μg streptomycin per ml. Duplicate sets to which mycostatin was also added at a concentration of 50 μg per ml medium, were tested as well. The culture tubes (16 by 150 mm) contained 1 ml medium, and each was inoculated with one or two female worms; the gas phase was 5% CO_2 in air. The tubes were rotated in drums at 12 revolutions per hour. Duplicate sets of all media tested were prepared; following inoculation of the cultures one set was incubated at 25 C, the other at 37 C. At the end of 24 hr, the worms were removed from the tubes, and all cultures of deposited eggs were then incubated at 25 C to permit normal embryo-

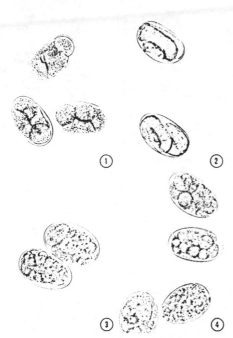

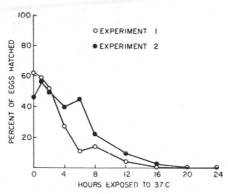

FIGURE 5. Percentage hatch of *N. dubius* eggs dissected from worm uteri and exposed for various time intervals to a temperature of 37 C, following which the eggs were incubated at 25 C. Approximately 3 hr should be added to each of these time intervals to take into account the dissection of worms and the preparation of the egg inoculum at 37 C.

FIGURES 1–4. *Nematospiroides dubius* eggs. × 220. **1**. In early cleavage 1 hr after being deposited at 25 C. **2**. Deposited and incubated in vitro at 25 C for 24 hr, containing normal, late vermiform embryos. **3**, **4**. Deposited and incubated at 37 C in vitro for 24 hr. Note abnormal cleavage patterns and termination of development at mid- to late blastula.

nation. Cultures were examined to determine the degree of embryo development and hatching after 24, 48, and 72 hr of incubation.

Preparation of cultures with eggs dissected from worms: Mature females were transferred to a sterile Petri dish containing a modified Krebs-Ringer-"Tris" solution (Sommerville and Weinstein, 1967) supplemented with 200 units penicillin and 100 μg streptomycin per ml (KRTA solution). The uterus of each worm was dissected with sterile, fine-pointed tungsten needles (Brady, 1965) under a dissecting microscope. Liberated eggs were removed by Pasteur pipette, pooled, and washed three times in the KRTA solution. Plastic flasks (tissue culture flask with screw cap, 30 ml sterile disposable plastic flask; Falcon Plastics, Division of B-D Laboratories, Inc.) containing 1.5 ml KRTA solution to a depth of about 1 mm, were then inoculated with approximately 100 eggs each. The flasks were incubated at specific temperature and time intervals described in the experiments below. The time from the necropsy of the host to the inoculation of the flasks was approximately 3 hr.

Sterility tests: All media, and cultures at termination were tested in thioglycollate broth for microbial contaminants.

RESULTS

1. *The development of deposited eggs at 25 C and 37 C.* Females in the various media tested, deposited between 130 to 250 eggs in early cleavage during the first 24 hr of incubation. In those cultures in which both egg deposition and subsequent incubation occurred at 25 C, approximately 40 to 70% of the eggs completed normal embryonation (Figs. 1, 2) and hatched. However, in the cultures in which egg deposition occurred at 37 C during a 24-hr period, followed by subsequent incubation at 25 C, none of the embryos developed beyond the late blastula (Figs. 3, 4). No marked differences were noted in the percentage of eggs that hatched in the various media tested. Mycostatin had no observable effect on embryonation.

In these studies, eggs were deposited by the worms during the entire 24-hr culture period, though most were laid in the initial portion. In order to achieve a more uniform experimental condition, all cultures prepared in the studies described below were inoculated with eggs dissected from the uteri of worms (see Methods).

2. *The development of eggs exposed to 37 C*

TABLE I. *Effect of temperature upon embryonation and hatching of N. dubius eggs incubated under nitrogen for 8 hr, followed by incubation in air.*

Culture treatment*	Percent of eggs hatched**
1. Gassed with N_2, incubated at 37 C for 8 hr, N_2 replaced by air and culture incubated at 25 C.	20.0; 20.0
2. Incubated in air at 37 C for 8 hr, followed by incubation in air at 25 C.	20.0
3. Gased with N_2, incubated at 25 C for 8 hr, N_2 replaced by air and culture incubated at 25 C.	47.4; 44.9
4. Incubated in air at 25 C.	52.5

* Duplicate cultures in 1 and 3; single cultures in 2 and 4.
** Evaluation made after 72 hr incubation.

TABLE II. *Effect of continuous exposure to 37 C upon N. dubius eggs in different stages of development.*

Stage of development of eggs when exposed to 37 C	Hours exposed to 37 C	Percent of eggs hatched*	
		Experiment 1	Experiment 2
Early cleavage	0**	50.0	50.4
Early cleavage	72	0	0
Late blastula	67	0	0
Tadpole "C"-shape	59	0	0
Early vermiform	54	0	0

* Evaluation made after 72 hr incubation.
** Incubated at 25 C only.

for various durations. Immediately after inoculation of a group of 10 flasks, the first was removed (zero hour) from the 37 C warmroom, and incubated at 25 C. Thereafter the remaining flasks were removed one at a time at 1, 2, 4, 6, 8, 12, 16, 20, and 24 hr after inoculation and, similarly, incubated at 25 C. Eggs in all cultures were examined at 24, 48, and 72 hr to determine the stage of embryonation attained. At the latter period, the stage of embryonation of each egg was recorded, and the percentage of hatched larvae was determined. The results of two such experiments are shown in Figure 5.

Following an exposure to 37 C of only 4 to 8 hr, the proportion of eggs that hatched decreased. Only a few eggs hatched after a 16-hr exposure, and none after 20 hr. The great majority of eggs exposed to 37 C for more than 16 hr died in the early stages of embryonation, none progressing beyond the late blastula.

3. *Effect of 37 C on eggs under nitrogen.* The above experiments indicated that exposure to a temperature of 37 C for relatively few hours was inimical to the development of eggs of *N. dubius.* These cultures were prepared with a gas phase of air. The eggs in vivo, however, must traverse the cecum and large intestine before being eliminated in the feces, and presumably are in an anaerobic environment for a considerable proportion of the time that they are in the host. An attempt was made, therefore, to determine whether an anaerobic atmosphere would modify and reduce the injurious temperature effect.

Egg cultures were prepared in plastic flasks, and were subjected to different conditions of gas phase and temperature (Table I). In those flasks with nitrogen as the gas phase, the individual flask was first flushed with nitrogen for 3 min, placed with loose cap into a gastight jar, which was then flushed with a stream of nitrogen for 30 min and sealed. Following the incubation period in nitrogen at either 25 or 37 C, the flask was removed from the jar, and gassed with a strong flow of air for 3 min. Incubation was then completed in air at 25 C.

The results of a typical experiment are presented in Table I. It is evident that the nitrogen atmosphere did not prevent injury at 37 C; decrease of the percent of eggs hatched was comparable to that of the culture incubated in air at the same temperature. Although the final percent of hatching was comparable in both of these sets of cultures, there was an indication that the rate of hatching was faster in air. At 25 C, the exposure of eggs to nitrogen for 8 hr had no apparent influence on subsequent development; no significant difference in the percent of eggs hatched was obtained between the nitrogen-treated eggs and those incubated with a gas phase of air.

4. *Effect of 37 C on eggs in different stages of embryonation.* To determine whether eggs in various stages of embryonation might show differences in susceptibility to injury by a temperature of 37 C, two types of experiments were performed (Tables II, III).

In the first, eggs dissected from mature worms at room temperature were inoculated into flasks, which were then incubated at 25 C for various intervals of time. When more than 50% of the eggs in a particular flask developed

TABLE III. *Effect of 8-hr exposure to 37 C upon N. dubius eggs in different stages of development, followed by incubation at 25 C.*

Stage of development of eggs when exposed to 37 C	Hours exposed to 37 C	Percent of eggs hatched*	
		Experiment 1	Experiment 2
Early cleavage	0	51.2	59.8
Early cleavage	8**	7.8	17.5
Late blastula	8**	0	0
Tadpole "C"-shape	8**	0	0
Early vermiform	8**	0	0

* Evaluation made after 72 hr incubation.
** After exposure to 37 C for 8 hr, cultures were reincubated at 25 C.

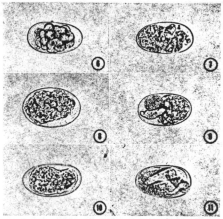

FIGURES 6–11. Abnormal development of *Nematospiroides dubius* eggs exposed to 37 C. All photomicrographs taken at the end of the total incubation period. × 220. **6.** In early cleavage; exposed to 37 C for 72 hr. **7.** In late blastula; exposed to 37 C for 8 hr and then incubated at 25 C for 59 hr. **8, 9.** In late blastula; exposed to 37 C for 67 hr. **10.** In tadpole "C"-shape; exposed to 37 C for 59 hr. **11.** With vermiform embryo; exposed to 37 C for 8 hr, and then incubated at 25 C for 46 hr.

either to early cleavage, blastula, tadpole "C"-shape, or early vermiform, i.e., 0, 5, 13, and 18 hr of incubation, respectively, each flask was incubated at 37 C and maintained at this temperature for 54 to 72 hr. For comparison with these, one flask with eggs in early cleavage was incubated at 25 C. The results of two separate experiments are presented in Table II. None of the eggs incubated at 37 C hatched, regardless of the stage of development, while approximately 50% of those eggs incubated at 25 C hatched.

In the second series, flasks of eggs representing the four different stages of embryonation described above, were placed at 37 C for 8 hr, after which the flasks were removed and reincubated at 25 C. Only some "early cleavage" eggs embryonated completely and hatched, whereas none of the eggs in later stages of development did so (Table III). Embryonation ceased in most of the eggs when they were placed at 37 C.

Figures 6 to 11 show the cleavage abnormalities and malformations that appeared in many of the eggs subjected to 37 C in the two types of experiments described above.

Microbial growth was absent in all cultures as judged by the sterility tests.

DISCUSSION

Nematodes demonstrate varying degrees of success in coping with the temperature "barrier" presented by warm-blooded hosts, and the existence of a differential temperature sensitivity among the stages in the life cycle of an organism is not unusual. This study has focused on the effects of temperature on the embryogenesis of the egg of a parasitic nematode. The maximum degree of embryonation attained by an egg within a warm-blooded host varies considerably among the nematode groups, ranging from eggs passed from the host as uncleaved fertilized ova (*Ascaris*), to those that complete embryonation and hatch in the host as first stage larvae (*Strongyloides stercoralis*). Within this spectrum are found eggs, such as those of trichostrongylids and strongylids, that undergo two or three cleavages before being passed from the body of the host. Cleavage, which in these cases is dependent upon the presence of oxygen, appears to be effectively interrupted as the egg passes through the anaerobic environment of the cecum and large intestine. The present findings in some instances, based on in vitro studies, indicate that temperature may also profoundly affect embryonic cell divisions.

Eggs of various strongylids and trichostrongylids have been reported to be capable of embryonating normally in vitro at 37 C and hatching; for example *Ancylostoma caninum* (Komiya, Yasuraoka, and Sato, 1956) and *A. duodenale* (Yasuraoka, Hosaka, and Ogawa,

1960) studied under axenic conditions, and *Haemonchus contortus* (Silverman and Campbell, 1958) in feces culture. The upper limit for development and hatching of eggs of three geographic variants of *H. contortus*, incubated in dialysis tubing, was found to range between 36 to 41 C (Crofton, Whitlock, and Glazer, 1965). However, investigations conducted with other trichostrongylids have demonstrated a decreased proportion of eggs hatched, when embryonated at 35 or 40 C (*Nippostrongylus brasiliensis* (Luttermoser, 1937); *Trichostrongylus retortaeformis* (Gupta, 1961); *Trichostrongylus colubriformis* (Anderson, Wang, and Levine, 1966; Wang, 1967)). These experiments were performed with eggs associated with bacteria, and possible injurious effects of microbial metabolism on embryonic development, particularly at the higher temperatures, could not be evaluated. The present studies, however, performed under axenic conditions provide direct evidence that a relatively short period of elevated temperature alone can be responsible for the injury.

Eggs deposited by *N. dubius* during a 24-hr culture period at 37 C, and subsequently incubated at 25 C did not complete embryonation. Various durations of exposure of eggs in early cleavage to 37 C indicated that as little as 4 to 8 hr reduced the proportion of eggs hatched, and that eggs exposed for more than 20 hr did not progress beyond the late blastula. Further studies of embryos at various stages of development ranging from early cleavage to late vermiform revealed in all a sensitivity to such an elevated temperature. However, a differential response was demonstrated in that embryos in relatively early cleavage were less affected by an 8-hr exposure to 37 C than were those in later stages of development. In more absolute terms, to all these figures should be added approximately the 3 hr at 37 C, required for the dissection of the worms and the preparation of the egg inoculum for the cultures.

In vivo, *N. dubius* deposits eggs into the lumen of the small intestine that are approximately in the 4- to 8-cell stage (Fig. 1). The average amount of time the eggs then remain in the mouse at a temperature of approximately 37 C before being eliminated in the feces is not precisely known. However, it may not average more than 2 or 3 hr. The time required for food to traverse the *entire* length of the digestive tract of the mouse (mouth to anus) has been reported to range from 2.5 to 8 hr, depending upon the amount of roughage in the diet; the one with the more "normal" roughage content had the shortest mobility time (DeWitt and Weinstein, 1964). In view of this, it is significant that under the test conditions used in the present study, eggs in early cleavage (dissected from the uterus) did not exhibit detectable heat injury prior to 4 hr (actually approximately 7 hr) exposure to 37 C. In addition, the early cleavage stage was relatively less sensitive to heat injury than the later stages of embryonation. This appears to be physiological adaptation in the life cycle of the parasite that has important survival value, yet the time-safety factor that exists does not seem to be very great. It is probable that a nematode such as *N. dubius*, even if it succeeded in developing to maturity in a warm-blooded host other than a mouse, would have its eggs injured or killed by heat-shock if the intestinal emptying time of the host exceeded the time-temperature tolerance relation of the egg.

Lacking comparative experimental data, however, it may be premature to attempt to extend in vitro findings to the in vivo situation. Although it would appear likely that *N. dubius* eggs deposited in the intestine would begin to exhibit heat shock if they were not evacuated from the host within a relatively few hours, it is conceivable that the environment of the cecum and large intestine could actually afford some type of protection to the egg. Incubating the eggs under nitrogen in a preliminary attempt to simulate the anaerobic condition characteristic of these portions of the alimentary canal, however, was of no consequence. Another factor that may have a bearing on the in vivo relation is the circadian temperature rhythm of the host. There is a difference of approximately 2 degrees between the lowest and highest mean rectal temperature of the mouse measured along a 24-hr scale, the minimum value approaching 35 C (Haus, Lakatua, and Halberg, 1967). This latter value is also 2 degrees lower than the temperature at which the cultures were incubated in the present study. It is possible that, at this lower tem-

64

perature, the *N. dubius* egg would exhibit a different time-temperature tolerance curve. Circadian defecation rhythms may similarly play a role.

The results of the present study have a direct bearing on the cultivation through successive generations of parasitic nematodes such as *N. dubius*. It is obvious that eggs deposited in culture will have to be harvested within a few hours, if uninjured embryos are to be obtained for the continuation of the free-living phase of the cycle. It appears highly probable that, failure to accomplish this accounted for the death of the eggs in early embryogenesis obtained in the previous study on cultivation of *N. dubius* (Sommerville and Weinstein, 1964). It is possible, of course, that the eggs in cleavage observed in that study actually were not fertile. Leland (1965, 1968), who similarly reported eggs in early cleavage in cultures of *Hyostrongylus rubidus* and *Cooperia oncophora*, considered the additional possibility that, in the case of *H. rubidus* cleavage might have been due to parthenogenetic development. True parthenogenesis, however, is rare among nematodes (Walton, 1940), and further study is needed to resolve this problem. Nevertheless, it is clear from the present investigation that *N. dubius* eggs of proven fertility are injured by incubation in culture for several hours at 37 C. To circumvent this problem, a modification of a device such as the cold trap developed by Tiner (1966) might be of value in harvesting and storing eggs as they are produced in vitro.

A somewhat puzzling and unexplained finding for all the cultures incubated at 25 C was the failure of a certain proportion of apparently fertile eggs to embryonate. Similar results have been reported for morph variants of *H. contortus cayugensis* (Crofton, Whitlock, and Glazer, 1965; Glazer, Crofton, and Whitlock, 1967). Based on dye penetration studies using acridine orange, Whitlock and his colleagues (pers. comm.) have determined that failure of a proportion of apparently fertile eggs of *H. contortus cayugensis* to embryonate is related to the fact that these eggs have different membrane permeability characteristics compared to mature eggs in the uterus.

The mechanism of heat injury cannot be determined from the experiments reported

here, but it is apparent that the normal pattern of cell cleavage was interrupted or severely distorted, resulting in malformed embryos. It is of interest in this regard that an analysis of the effects of heat-shock on the free-living nematode, *Caenorhabditis elegans,* has revealed the importance of nuclear injury (Brun, 1955; Nigon and Brun, 1967). This species reproduces normally when grown between 10 to 18 C. The relatively elevated temperature of 25 C induces in oogenesis, numerous chromosomal anomalies, fusions of the nuclei or irregular multiplications that are conducive to the formation of polyploid oocytes. At 31 to 32 C, the further evolution of the oocytes is blocked almost completely, the cells developing extremely slowly and abnormally. Fatt (1967) has further demonstrated that *C. elegans* subjected to heat-shock at 27 C in distilled water appears to need extra vitamins and salts (in addition to those provided by the liver extract of the axenic medium used for cultivation) to reproduce, giving some clue as to factors involved in protection against heat-shock.

LITERATURE CITED

ANDERSON, F. L., G.-T. WANG, AND N. D. LEVINE. 1966. Effect of temperature on survival of the free-living stages of *Trichostrongylus colubriformis.* J. Parasit. **52**: 713–721.

BRADY, J. 1965. A simple technique for making very fine, durable dissecting needles by sharpening tungsten wire electrolytically. Bull. Wld. Hlth. Organ. **32**: 143–144.

BRUN, J. 1955. Évolution de la prophase méiotique chez *Caenorhabditis elegans* Maupas, 1900, sous l'influence de température élevées. Bull. Biol. **89**: 326–346.

CROFTON, H. D., J. H. WHITLOCK, AND R. A. GLAZER. 1965. Ecology and biological plasticity of sheep nematodes. II. Genetic x environmental plasticity in *Haemonchus contortus* (Rud. 1803). Cornell Vet. **55**: 251–258.

DEWITT, W. B., AND P. P. WEINSTEIN. 1964. Elimination of intestinal helminths of mice by feeding purified diets. J. Parasit. **50**: 429–434.

FATT, H. V. 1967. Nutritional requirements for reproduction of a temperature sensitive nematode, reared in axenic culture. Proc. Soc. Exp. Biol. Med. **124**: 897–903.

GLAZER, R., H. D. CROFTON, AND J. H. WHITLOCK. 1967. Differential hatching of eggs from morph variants of *Haemonchus contortus cayugensis* (Nematoda, Trichostrongylidae). Cornell Vet. **57**: 194–200.

GUPTA, S. P. 1961. The effects of temperature

on the survival and development of the free-living stages of *Trichostrongylus retortaeformis* Zeder (Nematoda). Can. J. Zool. **39**: 47–53.

HAUS, E., D. LAKATUA, AND F. HALBERG. 1967. The internal timing of several circadian rhythms in the blinded mouse. Exp. Med. Surg. **25**: 7–45.

KOMIYA, Y., K. YASURAOKA, AND A. SATO. 1956. Survival of *Ancylostoma caninum* in vitro. Jap. J. Med. Sci. Biol. **9**: 283–292.

LELAND, S. E., JR. 1965. *Hyostrongylus rubidus*. In vitro cultivation of the parasitic stages including the production and development of eggs through five cleavages: A preliminary report. J. Parasit. **51** (no. 2, sect. 2): 47.

———. 1968. In vitro egg production of *Cooperia oncophora*. J. Parasit. **54**: 136.

LUTTERMOSER, G. 1937. Factors influencing the development and viability of the eggs of *Nippostrongylus muris*. J. Parasit. **23**: 539–540.

McQUILKIN, W. T., V. J. EVANS, AND W. R. EARLE. 1957. The adaptation of additional lines of NCTC clone 929 (strain L) cells to chemically-defined protein-free medium NCTC 109. J. Natl. Cancer Inst. **19**: 885–907.

NIGON, V., AND J. BRUN. 1967. Génétique et évolution des nématodes libres. Perspectives tirées de l'étude de *Caenorhabditis elegans*. Experientia **23**: 161–170.

SILVERMAN, P. H., AND J. A. CAMPBELL. 1958. Studies on parasitic worms of sheep in Scotland. I. Embryonic and larval development of *Haemonchus contortus* at constant conditions. Parasitology **49**: 23–38.

SOMMERVILLE, R. I., AND P. P. WEINSTEIN. 1964. Reproductive behavior of *Nematospiroides dubius* in vivo and in vitro. J. Parasit. **50**: 401–409.

———, AND ———. 1967. The in vitro cultivation of *Nippostrongylus brasiliensis* from the late fourth stage. J. Parasit. **53**: 116–125.

TINER, J. D. 1966. Collection and storage of axenic inoculum of plant parasitic nematodes in the laboratory. Ann. N. Y. Acad. Sci. **139**: 111–123.

WALTON, A. C. 1940. Gametogenesis, p. 205–215. *In* B. G. Chitwood and M. B. Chitwood, An Introduction to Nematology, Leader Press, Babylon, New York.

WANG, G. T. 1967. Effect of temperature and cultural methods on development of the free-living stages of *Trichostrongylus colubriformis*. Am. J. Vet. Res. **28**: 1085–1090.

WEINSTEIN, P. P., AND M. F. JONES. 1959. Development in vitro of some parasitic nematodes of vertebrates. Ann. N. Y. Acad. Sci. **77**: 137–162.

YASURAOKA, K., Y. HOSAKA, AND K. OGAWA. 1960. Survival of *Ancylostoma duodenale* in vitro. Jap. J. Med. Sci. Biol. **13**: 207–212.

THE EFFECT OF GROWTH INHIBITORS ON POSTEMBRYONIC DEVELOPMENT IN THE FREE-LIVING NEMATODE, *PANGRELLUS SILUSIAE**

J. PASTERNAK and M. R. SAMOILOFF

INTRODUCTION

POSTEMBRYONIC growth of the free-living nematode *Panagrellus silusiae* can be synchronized by isolating a homogeneous population of the first free-swimming stage (Samoiloff & Pasternak, 1968). This system facilitated the demonstration that certain enzymic proteins vary in a precise manner during maturation (Chow & Pasternak, 1969) and also enabled the examination of the ultrastructure during the moulting cycles (Samoiloff & Pasternak, 1969). The study of postembryonic development is of interest because several nematodes have the unique property of retaining a constant number of somatic cells after completion of embryogenesis despite the continuation of extensive postpartum growth (Martini, 1908; Pai, 1928; Wessing, 1953). The reproductive system is the principal exception to cell constancy although some workers have reported that intestinal and hypodermal cells may have a meagre capacity to proliferate during the growth period (Moorthy, 1938; Wessing, 1953). Growth, therefore, is due to an enlargement of cell size and not to an increase in cell number.

Moulting and the formation of the reproductive system are two major morphogenetic processes during postembryonic development. Nematodes undergo four

* This work was supported by a National Research Council of Canada grant (A-3491) and by Research Corporation, Brown Hazen Fund Project.

moults. In *P. silusiae* the first moult occurs *in utero* and the last three occur at intervals during the postpartum growth period. The elaboration of a functional reproductive system follows a definite sequential pattern. Synchronous development should, therefore, provide an opportunity for analyzing the relationships between morphological events and the biochemical changes that take place.

The experiments to be described were undertaken to characterize the patterns of DNA, RNA and protein synthesis in the whole organism during postembryonic development in synchronized cultures and to study the effect of presumed inhibitors of DNA, RNA and protein synthesis on growth and macromolecular biosynthesis.

MATERIALS AND METHODS

Nematode culture

Culture conditions and the method of obtaining synchronous postembryonic growth of the nematode *Panagrellus silusiae* have been described in previous reports from this laboratory (Samoiloff & Pasternak, 1968, 1969; Chow & Pasternak, 1969). In the present study the nematodes used were from strain C-15 of *P. silusiae*. This strain was derived from the original culture obtained from Dr. A. C. Coomans (Gent, Rijksuniversitat) after fifteen consecutive generations of brother–sister matings. Synchronous growth was monitored by measuring the length of nematodes ($N \geqslant 50$) stained and fixed with 0·0025% cotton blue-lactophenol (Goodey, 1957).

Labeling procedures

Radioactive labeling was started by the addition of 100 μl of radioactive thymidine, uridine or leucine into a depression containing 0·75 ml of fresh medium and about 500 nematodes.

Thymidine methyl-³H (> 15 c/mM), uridine-5-³H (> 20 c/mM) and L-leucine-4,5-³H (~ 40 c/mM) were purchased from New England Nuclear Corporation. At the end of the desired incubation period, 200 nematodes were removed from the radioactive medium and passed through six successive depressions containing nonradioactive medium. Immediately after the last wash the nematodes were pipetted into cold 6% TCA. The nematodes labeled with thymidine-³H or uridine-³H were stored at 0°C overnight. The nematodes treated with tritiated leucine were hydrolyzed for 20 min at 93°C and then stored at 0°C overnight.

The labeled nematodes were collected on glass fiber filters (Reeve-Angel). The filters were rinsed three times with 5 ml of cold 5% TCA and once with 15 ml of cold saline, dried and placed in scintillation vials. Radioactivity was determined in a Packard Tri-Carb scintillation counter. The vials contained 10 ml of toluene-PPO-dimethyl POPOP scintillation mixture (5 g PPO and 0·5 g dimethyl POPOP in 1 l. toluene).

To test the specificity of the incorporation of the radioactive compounds, samples of synchronously growing nematodes were taken at various times after the addition of tritiated precursor (final conc. ~ 35 μc/ml), and washed five times by centrifugation (450 g) in 0·15 M NaCl. The final sample was suspended in 1 ml of 0·15 M NaCl and frozen. After thawing, the nematodes were disrupted by sonication at 25 W (Branson Cell Disruptor, Heat Systems Co.) for 3 min. The suspension was divided into two equal amounts and subsequent treatment depended upon the labeled compound used. With the suspension that had thymidine-³H as the label, one portion was put into an equal volume of cold 10% TCA and the other portion was incubated with 50 μg/ml DNase (RNase free, Worthington) at 37°C for 1 hr. The uridine-³H labeled suspension was processed analogously except that 50 μg/ml RNase (3 × cryst., Worthington) replaced the DNase. The enzymic digestions were terminated by the addition of an equal volume of cold 10% TCA. With the suspension containing leucine-³H labeled material, one part was hydrolyzed at 93°C for 20 min in

5% TCA (final conc.) and the other part was digested with 60 μg/ml of self-digested pronase (CalBiochem) at 37°C for 3 hr and then hydrolyzed in 5% TCA. The suspensions were stored at 0°C overnight. The unfilterable material was collected on glass fiber filters. Radioactivity was determined by scintillation counting.

Chemicals

Actinomycin D, nalidixic acid and phleomycin were the generous gifts of Dr. W. Dorion, Merck, Sharp and Dohme of Canada, Ltd., Dr. C. W. Birkett of Winthrop Laboratories, and Dr. A. Gourevitch of Bristol Laboratories, respectively. Chloramphenicol was from Parke, Davis and Co. Puromycin was obtained from Nutritional Biochemicals Corporation. Actidione (cycloheximide) was purchased from both Nutritional Biochemicals Corporation and Sigma Chemical Co. The same results were obtained with actidione from either source. Phenethyl alcohol and hydroxyurea were from the Aldrich Chemical Co.

Autoradiography

To determine the sites of radioactivity, whole animals or sectioned material (0·45 μ) were processed for autoradiography according to the method of Caro (1964).

RESULTS

Synchronous postembryonic growth

The growth curve of the C-15 strain of *P. silusiae* is shown in Fig. 1. The rate and degree of synchronous growth may vary from experiment to experiment, but the same pattern is observed. When three or more replicate samples are grown simultaneously from the same source of juvenile worms, there is little variation in the rate of growth.

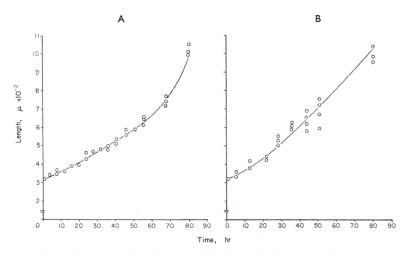

FIG. 1. Growth curves of *P. silusiae* strain C-15. Each open circle represents at least one sample ($N \geqslant 50$). When more than one replicate sample had the same mean length one symbol was used. A. One experiment in triplicate. B. One experiment in quintuplicate.

Effect of inhibitors on synchronous growth

Initially, the effect of at least six concentrations of each inhibitor was tested to establish a dose level that was not lethal to the nematodes for 7 days in continuous culture. The highest concentration of each inhibitor fulfilling this criterion was not chosen for further study; instead, the concentrations effective below the critical level were used in order to lessen non-specific metabolic effects. The effect of selected concentrations of the inhibitors on postembryonic growth is illustrated in Figs. 2 and 3. Actinomycin D, actidione, chloramphenicol, puromycin (200 μg/ml) and phleomycin (200 μg/ml) inhibited growth, whereas hydroxyurea (400 μg/ml) and phenethyl alcohol had a slightly inhibitory effect on growth. Nalidixic acid (200 μg/ml) had no effect on the growth rate.

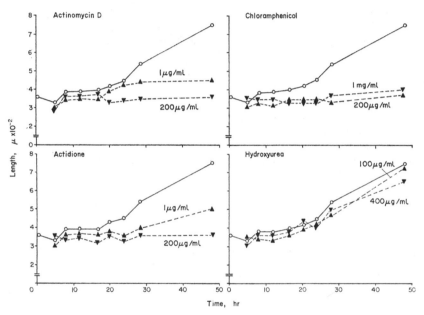

Fig. 2. The effect of two concentrations of actinomycin D. chloramphenicol, actidione and hydroxyurea on postembryonic growth. The substances were added at zero time. O----O, Untreated control; ▲----▲, treated with lower concentration; ▼----▼, treated with higher concentrations.

When nematodes are returned to antibiotic-free medium within 36 hr after continuous treatment with inhibiting concentrations, only partial recovery is observed. The synchronous growth is disrupted. Some of the treated nematodes remain permanently inhibited, others resumed the normal growth rate after a lag period. The number of nematodes that recover from 36 hr of treatment and the duration of the lag period is extremely variable from experiment to experiment. Only 10–20 per cent of the nematodes become sexually competent adults. If nematodes that are maintained in growth-inhibiting concentrations for more than 36 hr are transferred to fresh medium without the inhibitor, recovery is rare and less than 1 per cent of the nematodes reach sexual maturity.

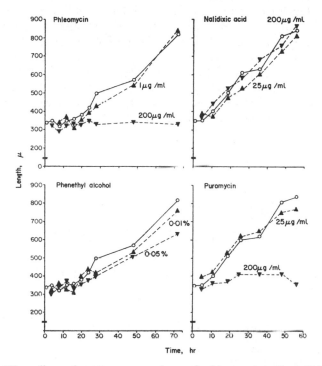

FIG. 3. The effect of two concentrations of phleomycin, phenethyl alcohol, puromycin and nalidixic acid on growth. The substances were added at zero time. O−−−−O, Control; ▲−−−−▲, treated with lower concentration; ▼−−−−▼, treated with higher concentration.

Actinomycin D (20 μg/ml), actidione (75 μg/ml) and puromycin (200 μg/ml) were added to synchronously growing populations at various times during the postembryonic period. Further growth was inhibited, indicating that there are no antibiotic resistant periods during maturation (Fig. 4).

DNA, RNA and protein synthesis

DNA, RNA and protein synthesis during the postembryonic growth period was measured by continuous labeling with thymidine-³H, uridine-³H or leucine-³H, respectively. The specificity of incorporation of each precursor was ascertained (Table 1). The data show that each precursor acted according to expectation. The long incubation periods did not alter the degree of precursor specificity.

71

Autoradiographs of nematodes removed from thymidine-³H, uridine-³H or leucine-³H at various times during the maturation period revealed that the lumen was virtually free of silver grains while the body tissues were heavily labeled. The

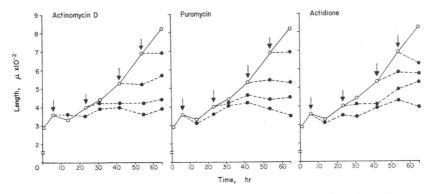

FIG. 4. The effect of the addition of actinomycin D, puromycin and actidione at different times during postembryonic growth. Arrows denote the time of addition of actinomycin D (20 μg/ml), puromycin (200 μg/ml) and actidione (100 μg/ml). Each inhibitor was added at about 14 hr and about 30 hr with comparable results. These data are omitted. O– – – –O, Control; ●– – – –●, treated.

TABLE 1—SPECIFICITY OF URIDINE-³H, THYMIDINE-³H AND LEUCINE-³H AS PRECURSORS FOR RNA, DNA AND PROTEIN

Time in label (hr)	Length (μ ± S.D.)	Counts/min		RNA (%)
		No RNase	After RNase	
Uridine-³H				
5	315·05 ± 19·11	15,651	4460	71
12	320·44 ± 27·07	6298	2301	63
50	540·77 ± 133·21	7624	664	91
76·5	972·62 ± 132·14	9119	1080	89
		No DNase	After DNase	DNA (%)
Thymidine-³H				
12	296·67 ± 15·81	1558	983	69
28	350·91 ± 17·18	2822	1212	58
50	533·87 ± 138·79	13,535	2229	84
76·5	969·22 ± 141·21	55,567	8921	84
		No Pronase	After Pronase	Protein (%)
Leucine-³H				
5	321·80 ± 20·45	39,633	5607	86
28	396·67 ± 76·87	66,780	10,703	84
50	505·00 ± 36·79	36,780	7890	79
76·5	934·83 ± 101·21	104,671	9040	91

amount of label in the lumen was equivalent to the number of silver grains per unit area of background.

The rate of incorporation of isotopic precursors was used as a measure of the gross rates of DNA, RNA and protein synthesis. Although the results are given for one representative experiment, the individual experiments were repeated at least four separate times with similar results. The exceptions are noted.

The uptake of thymidine-^{3}H into DNA was continuous during postembryonic growth (Fig. 5). In two out of ten similarly designed experiments the pattern of incorporation of precursor material showed a significant drop in the net accumulation of labeled DNA at 40 hr after incubation. Such discrepancies probably reflect an occasional perturbation of growth in replicate samples.

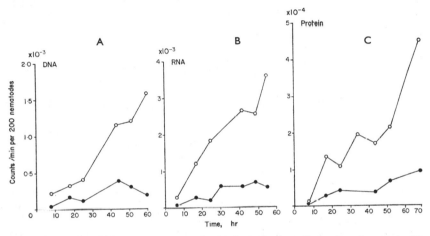

FIG. 5. DNA, RNA and protein synthesis and the effect of actinomycin D (15 μg/ml) during postembryonic growth. A. DNA synthesis was measured by the incorporation of thymidine-^{3}H into an acid-insoluble fraction. B. RNA synthesis was measured as in A using uridine-^{3}H as precursor. C. Protein synthesis was measured by leucine-^{3}H incorporation into a hot acid precipitable fraction. The radioactive precursor and actinomycin D were added at zero time. O– – – –O, Control; ●– – – –●, actinomycin D-treated.

RNA synthesis continued in a linear manner during the growth period. The accumulation of leucine-^{3}H in protein is usually continuous during postembryonic growth (Fig. 5). In four out of ten experiments protein synthesis followed a biphasic pattern with an acceleration at the rate of incorporation occurring about 50–60 hr after initiation of synchrony.

Actinomycin D (15 μg/ml) inhibited all three types of macromolecular synthesis. DNA synthesis appeared slightly more susceptible to actidione inhibition than either RNA or protein synthesis. Hydroxyurea (500 μg/ml) causes only about 15 per cent inhibition of growth, but blocks RNA and DNA synthesis by about 50 per cent and markedly interferes with protein synthesis (Fig. 6).

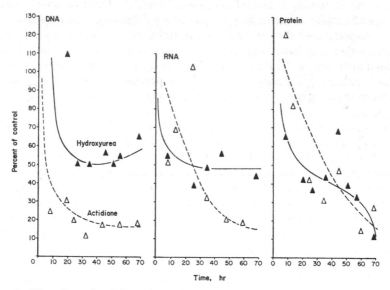

Fig. 6. The effect of actidione (75 μg/ml) and hydroxyurea (500 μg/ml) on DNA, RNA and protein synthesis during growth. The experimental design is the same as described in Fig. 5. △– – – –△, Actidione-treated; ▲– – – –▲, hydroxyurea-treated. The dotted and solid lines indicate trends.

DISCUSSION

A series of morphological, ultrastructural and metabolic changes are associated with the postembryonic development of nematodes. These changes occur in the presumed absence of cell division in all organ systems except for the reproductive system. As yet little information exists concerning macromolecular biosynthesis during the postembryonic growth period. The present experiments indicate that DNA, RNA and protein synthesis occur concomitantly during the growth period. The growth of the reproductive system is not continuous through postembryonic development in *P. silusiae*. Substantial cell proliferation begins when the nematodes are about 600 μ in length or about 50–60 hr of growth after isolation of L2 nematodes (unpublished results). Therefore, the incorporation of thymidine-^{3}H into DNA, at least for the first 50 hr of synchronous growth, is not due solely to DNA synthesis in gonadal cells. Nonnenmacher-Godet & Dougherty (1964) have shown that in *Caenorhabditis briggsae* the nuclei and cytoplasm of the intestinal cells will take up thymidine-^{3}H into DNA. Using qualitative autoradiography, we have observed the accumulation of thymidine-^{3}H into DNase-removable material in the intestinal cells, the chordal regions of the hypodermis and the noncontractile portion of the muscle layer. To locate the sites of radioactivity more precisely, we are examining this problem utilizing high-resolution autoradiography.

The effects of several known inhibitors on growth and macromolecular synthesis were studied under conditions employing inhibitor concentrations that did not cause death within 7 days of continuous contact. Unfortunately, knowledge of the specific mode of action of each inhibitory substance is problematic. Moreover, an inspection of the extensive literature on growth inhibitors indicates that each compound may have secondary or tertiary sites of action, especially during prolonged incubations. The issue is further complicated by the fact that some of the inhibitors may not act the same way in both procaryotes and eucaryotes. Since the mode of action of each inhibitor is not precisely known, an explicit interpretation of the present results is not possible.

Actinomycin inhibits DNA-dependent RNA synthesis by binding to DNA (Reich & Goldberg, 1964) and blocks DNA synthesis directly (Prudhomme et al., 1968). The fact that actinomycin D inhibits nematode growth as well as each of the three kinds of macromolecular synthesis examined suggests that both DNA and RNA synthesis are blocked directly and the inhibition of protein synthesis is secondary. After the addition of actinomycin D at various times during the growth cycle, no residual growth occurs. This observation indicates that there is no build-up of cellular materials required for growth and that DNA and RNA synthesis, or both, are essential for growth.

Comparable results with two protein synthesis inhibitors, puromycin (Nathans, 1964) and actidione (Wettstein et al., 1964), demonstrate that the nematode must maintain protein synthesis for growth. Actidione is capable of blocking DNA, RNA and/or protein synthesis in different biological systems (Bennett et al., 1964; Ennis & Lubin, 1965; DeKloet, 1966; Ennis, 1966). The inhibition of nucleic acid and protein synthesis by actidione in nematodes (see Fig. 6), then, can be explained by the action of actidione on the three syntheses directly, or by a network of interrelationships of syntheses where one type of macromolecular production is dependent upon the antecedent synthesis of another macromolecule. Until more is known about the action of actidione in nematodes, we prefer the former explanation. Chloramphenicol effectively blocked nematode growth. The site of action of this compound is not known. Weisberger (1967) proposed that chloramphenicol hinders messenger RNA attachment to ribosomes in eucaryotes, although other workers have suggested that this compound can disrupt energy metabolism (Stoner et al., 1964; Godchaux & Herbert, 1966) and alter mitochondrial structure (Marchant & Smith, 1968; Firken & Linnane, 1969).

The group of compounds that can be nominally classified as DNA synthesis inhibitors gave extremely variable results. Nalidixic acid was without effect on growth. Cook et al. (1966) have shown that nalidixic acid degrades bacterial DNA in situ. It is possible that the protein covering of chromosomes of higher organisms affords sufficient protection for the DNA. Phleomycin at high concentrations effectively blocked nematode growth. In this case phleomycin may block RNA synthesis (Kihlman et al., 1967; Watanabe & August, 1968) as well as DNA synthesis (Tanaka et al., 1963).

Phenethyl alcohol only partially blocked growth. Originally phenethyl alcohol was described as a selective inhibitor of DNA synthesis (Berrah & Konetzka, 1962; Leach et al., 1964). More recently it has been proposed that phenethyl alcohol alters membrane structure (Lester, 1965; Silver & Wendt, 1967). Plagemann (1968) suggests that phenethyl alcohol may modify an overall growth control center. Since the mode of action of phenethyl alcohol is unsettled, we cannot offer any reasonable explanation of the observed slight inhibition of growth. Higher concentrations of phenethyl alcohol ($0 \cdot 1$–$0 \cdot 5\%$), however, are lethal to nematodes within a few hours. This observation suggests that phenethyl alcohol may act on a ubiquitous and vital cell component such as cell membranes. Hydroxyurea acts as a potent inhibitor of DNA synthesis (Young & Hodas, 1964; Sinclair, 1967). Recently, Rosenkranz et al. (1969) have shown that iso-hydroxyurea is an effective inhibitor of RNA and protein synthesis in vivo. They suggest that this compound could be a natural breakdown product of hydroxyurea. Hydroxyurea treatment has little significant effect on nematode growth, although the synthesis of DNA, RNA and protein are inhibited to an appreciable degree. This result does not contradict our findings that continual macromolecular synthesis is a requirement for growth. Microscopic inspection of the hydroxyurea-treated organisms revealed that gonad formation is severely inhibited or drastically altered. The decrease in incorporation of radioactive precursors into DNA, RNA and protein is due in major part to the specific disruption of normal reproductive system development. The patterns of macromolecular synthesis during the development of the reproductive system remain to be described.

REFERENCES

BENNETT L. L., JR., SMITHERS D. & WARD C. T. (1964) Inhibition of DNA synthesis in mammalian cells by actidione. Biochim. biophys. Acta 87, 60–69.
BERRAH G. & KONETZKA W. A. (1962) Selective and reversible inhibition of the synthesis of bacterial deoxyribonucleic acid by phenethyl alcohol. J. Bacteriol. 83, 738–744.
CARO L. G. (1964) High-resolution autoradiography. In Methods in Cell Physiology, Vol. I (Edited by PRESCOTT D. M.), pp. 327–363. Academic Press, New York.
CHOW H. H. & PASTERNAK J. (1969) Protein changes during maturation of the free-living nematode, Panagrellus silusiae. J. exp. Zool. 170, 77–84.
COOK T. M., DIETZ W. H. & GOSS W. A. (1966) Mechanism of action of nalidixic acid in Escherichia coli—IV. Effects on the stability of cellular constituents. J. Bacteriol. 91, 774–779.
DEKLOET S. R. (1966) The effect of cycloheximide on the synthesis of ribonucleic acid in Saccharomyces carlsbergensis. Biochem. J. 99, 566–581.
ENNIS H. L. (1966) Synthesis of ribonucleic acid in L cells during inhibition of protein synthesis by cycloheximide. Mol. Pharmacology 2, 543–557.
ENNIS H. L. & LUBIN M. (1965) Cycloheximide: Aspects of inhibition of protein synthesis in mammalian cells. Science 146, 1474–1476.
FIRKEN F. C. & LINNANE A. W. (1969) Biogenesis of mitochondria-8. The effect of chloramphenicol on regenerating rat liver. Expl Cell Res. 55, 68–76.
GODCHAUX W., III, & HERBERT E. (1966) The effect of chloramphenicol on intact erythroid cells. J. Mol. Biol. 21, 537–553.
GOODEY J. B. (1957) Laboratory Methods for Work with Plant and Soil Nematodes. Technical Bulletin No. 2. Her Majesty's Stationery Office, London.
KIHLMAN B. A., ODMARK G. & HARTLEY B. (1967) Studies of the effects of phleomycin on chromosome structure and nucleic acid synthesis in Vicia faba. Mutation Res. 4, 783–790.

LEACH F. R., BEST H., DAVIS E. M., SANDERS D. C. & GRIMLIN D. M. (1964) Effect of phenethyl alcohol on cell culture growth—I. Characterization of the effect. *Expl Cell Res.* **36**, 524–532.

LESTER G. (1965) Inhibition of growth, synthesis and permeability in *Neurospora crassa* by phenethyl alcohol. *J. Bacteriol.* **90**, 29–37.

MARCHANT R. & SMITH D. G. (1968) The effect of chloramphenicol on growth and mitochondrial structure of *Pythium ultimum*. *J. gen. Microbiol.* **50**, 391–397.

MARTINI E. (1908) Die Konstanz histologischer Elemente bei Nematoden nach Abschluss der Entwickelungsperiode. *Verh. anat. Ges., Jena* **32**, 132–134.

MOORTHY V. N. (1938) Observations on the life history of *Camallanus sweeti*. *J. Parasit.* **24**, 323–342.

NATHANS D. (1964) Inhibition of protein synthesis by puromycin. *Fedn Proc. Fedn Am. Socs. exp. Biol.* **23**, 984–989.

NONNENMACHER-GODET J. & DOUGHERTY E. C. (1964) Incorporation of tritiated thymidine in the cells of *Caenorhabditis briggsae* (Nematoda) reared in axenic culture. *J. Cell Biol.* **22**, 281–290.

PAI S. (1928) Die Phasen des Lebenscyclus der *Anguillula aceti* Ehrbg. und ihre experimentell-morphologischen Beeinflussung. *Z. wiss. Zool.* **131**, 293–344.

PLAGEMANN P. G. W. (1968) Phenethyl alcohol. Reversible inhibition of synthesis of macromolecules and disaggregation of polysomes in rat hepatoma cells. *Biochim. biophys. Acta* **155**, 202–218.

PRUDHOMME J. C., GILLOT S. & DAILLE J. (1968) Effets de l'actinomycine D sur la synthèse de l'ADN et de l'ARN dans la grande séricigène de *Bombyx mori* L. *Expl Cell Res.* **48**, 186–189.

REICH E. & GOLDBERG I. H. (1964) Actinomycin and nucleic acid function. In *Progress in Nucleic Acid Research and Molecular Biology* (Edited by DAVIDSON J. N. & COHN W. E.), Vol. 3, 183–234. Academic Press, New York.

ROSENKRANZ M. S., POLLACK R. D. & SCHMIDT R. M. (1969) Biologic effects of isohydroxyurea. *Cancer Res.* **29**, 209–218.

SAMOILOFF M. R. & PASTERNAK J. (1968) Nematode morphogenesis: Fine structure of the cuticle of each stage of the nematode, *Panagrellus silusiae* (de Man 1913) Goodey 1945. *Can J. Zool.* **46**, 1019–1022.

SAMOILOFF M. R. & PASTERNAK J. (1969) Nematode morphogenesis: Fine structure of the moulting cycles in *Panagrellus silusiae* (de Man 1913) Goodey 1945. *Can. J. Zool.* **47**, 639–644.

SILVER S. & WENDT L. (1967) Mechanism of action of phenethyl alcohol: Breakdown of the cellular permeability barrier. *J. Bacteriol.* **93**, 560–566.

SINCLAIR W. K. (1967) Hydroxyurea: Effects on Chinese hamster cells grown in culture. *Cancer Res.* **27**, 297–308.

STONER C. C., HODGES T. K. & HANSON J. B. (1964) Chloramphenicol as an inhibitor of energy-linked processes in maize mitochondria. *Nature, Lond.* **203**, 258–261.

TANAKA N., YAMAGUCHI H. & UMEZAWA H. (1963) Mechanism of action of phleomycin—I. Selective inhibition of the DNA synthesis in *E. coli* and in HeLa cells. *J. Antibiot. Tokyo Ser. A* **16**, 86–91.

WATANABE M. & AUGUST J. T. (1968) Replication of RNA bacteriophage R23—II. Inhibition of phage-specific RNA synthesis by phleomycin. *J. Mol. Biol.* **33**, 21–33.

WEISBERGER A. S. (1967) Inhibition of protein synthesis by chloramphenicol. *A. Rev. Med.* **18**, 483–494.

WESSING A. (1953) Histologische Studien zu den Problem der Zellkonstanz Untersuchungen an *Rhabditis anomala* P. Hertwig. *Zool. Jb. Abt. Anat. Ontog. Tiere* **73**, 69–102.

WETTSTEIN F. O., NOLL H. & PENMAN S. (1964) Effect of cycloheximide on ribosomal aggregates engaged in protein synthesis *in vitro*. *Biochim. biophys. Acta* **87**, 525–528.

YOUNG C. W. & HODAS S. (1964) Hydroxyurea: Inhibitory effect on DNA metabolism. *Science* 1172–1174.

Heme Requirement for Reproduction of a Free-Living Nematode

W. F. Hieb
E. L. R. Stokstad

Several species of free-living nematodes have been serially cultured in axenic media (*1*). Reproduction of these organisms occurs only when the chemically defined medium (*2*) is supplemented with tissue extracts such as those from liver or chick embryos (*1, 3*). Certain fractions of liver extract reportedly contain a single, biologically active protein (*4*) which possesses specific structural characteristics (*5*). However, this purified "growth factor" contains small amounts of lipids and nucleic acids in addition to protein (*4, 6*).

We have observed that intact *Escherichia coli* will support the indefinite culture of the small, free-living nematode *Caenorhabditis briggsae* in a buffer-salt medium, providing that sterols are added (*7*). However, if the bacterial cells are first autoclaved, the nematodes will not reproduce, even in the presence of sterols and defined medium. Thus, some essential growth component in the cells is destroyed by heat. We have recently obtained a fraction from heated lamb liver extract which will substi-

tute for this component. This material, obtained by chromatography of the extract on Sephadex G-100, yielded a red fraction (fraction A) which supported reproduction of the nematodes only if the defined medium contained autoclaved bacterial cells plus sterols (*8*). This liver fraction remains active after autoclaving for 8 minutes at 120°C. We now report that fraction A contains a hemeprotein and that the biological effect of this fraction can be duplicated by pure myoglobin, cytochrome c, hemoglobin, and hemin.

The heated liver extract was prepared by homogenizing approximately 50 g of lamb liver with an equal amount of water, by heating at 53°C for 6 minutes and by centrifuging at 39,000g for 30 minutes as described by Sayre *et al.* (*2*). The resulting supernatant solution is a deep crimson when fresh. The absorption spectrum of this solution is the same as that of the biologically active fraction A isolated from it. In both cases an absorption maximum occurs in the Soret region at about 415

79

Table 1. Assay of various supplements for their ability to support reproduction of *C. briggsae*. Compounds were added to a defined basal medium that contained sterols and autoclaved *E. coli* cells (see text). In experiment 1 the values in parentheses indicate the concentration of heated liver extract in percent by volume. Duplicate tubes containing 0.25 ml of medium were inoculated with three newly hatched larvae and incubated at 20°C. Larvae were observed daily and increments of growth were measured by an arbitrary grading system which was then converted to actual length. Generation time was estimated as the time at which the first newly hatched larvae appeared after the original organisms grew from a length of 200 μm to maturity. Final population was estimated after a period of 15 to 20 days. The population range is designated as $+$ to $+++++$, the maximum number of worms per tube observed under the most favorable conditions. A population of $++$ or more typically indicates that a second generation has appeared as a result of the maturation of the offspring of the original larvae.

Supplement	Concentration (μg/ml)	Growth rate (μm/day)	Generation time (days)	Final population
Experiment 1				
Heated liver extract	0	140	n.r.*	
(53°C for 6 minutes)	5 (0.16%)	190	5.0†	+
	10 (0.31%)	200	3.9	++++
	20 (0.63%)	180	3.9	+++++
Experiment 2				
Myoglobin	0	140	n.r.	
(not autoclaved)	25	200	6.3†	+
	50	190	3.5	++
	100	170	3.7	++++
Experiment 3				
Myoglobin (autoclaved	0	140	n.r.	
for 8 minutes at 120°C)	12.5	100	n.r.	
	25	140	9–11	+
	50	190	3.2	++++
Experiment 4				
Hemoglobin	0	160	n.r.	
	12.5	200	n.r.	
	25	270	2.6	++
	50	310	2.7	+++
	100	240	2.6	++++
Cytochrome c	25	260	2.7	+++
	50	240	3.1	++++
	100	230	3.3	+++++
Experiment 5				
Cytochrome c	0	150	n.r.	
β-Lactoglobulin	100	140	n.r.	
(not autoclaved)				
β-Lactoglobulin	100	80	n.r.	
(autoclaved)				
Bovine serum albumin	100	110	n.r.	
Soluble casein	100	80	n.r.	
(autoclaved)				
Experiment 6				
Hemin chloride	0	160	n.r.	
	1	170	7.7	+
	2	210	3.8	+
	4	210	3.1	++
	8	240	3.0	++++

* No reproduction.   † Individuals in only one tube reproduced.

nm, and satellite bands are present at 540 and 580 nm. The spectrum is characteristic of either oxymyoglobin or oxyhemoglobin. Based on the extinction coefficient at 418 nm, the concentration of hemeproteins in the original heated liver extract is approximately 3.0 mg/ml of extract or 6.0 mg/g (wet weight) of original liver.

There is evidence that the hemeprotein in fraction A and heated liver extract is myoglobin rather than hemoglobin or cytochrome c. Preparation of the extract at 53°C precipitates proteins such as hemoglobin to a large extent but not myoglobin (9); the chromatographic properties of fraction A on Sephadex correspond to those of a substance with a molecular weight similar to that of myoglobin rather than hemoglobin (8); fraction A is readily distinguished from cytochrome c by gel electrophoresis in a pH gradient.

The presence of hemeproteins suggested that this type of compound was implicated in the nematode "growth factor." The following information offers proof of this.

The assay of myoglobin and other materials for growth activity in C. briggsae is similar to that reported (3, 7). The chemically defined medium employed was originally reported by Sayre et al. (2) and modified by Buecher et al. (10). It consists of pure amino acids, glucose, water soluble vitamins, salts (including iron), and ribonucleotides (11). This mixture was further supplemented with sterols (50 μg/ml) in Tween 80 as described (7), and with autoclaved E. coli cells [0.3 mg/ml (dry weight)], which altogether comprise the basal medium. Axenic larvae of C. briggsae (usually three) were inoculated into a 10- by 75-mm culture tube that contained 0.25 ml of medium (in duplicate) and were incubated at 20°C. The effectiveness of the medium was determined by observing at intervals of 1 day or less the growth in length of larvae and reproduction of adult individuals.

In the basal medium alone, larvae grow to the size of advanced fourth stage larvae or small adults (650 to 850 μm long) in about 4 days, but they do not reproduce. As little as 0.31 percent (by volume) of heated liver extract (equivalent to 1.6 mg of original liver per milliliter of medium) supported extensive reproduction of the larvae (Table 1). Doubling the concentration of heated liver extract had little

effect, whereas 0.16 percent proved to be inadequate.

As long as autoclaved bacterial cells are added, purified myoglobin (equine heart, Pierce Biochemicals, Rockford, Ill.) can replace heated liver extract as a growth requirement (Table 1). Myoglobin was dissolved in water at a concentration of 5 mg/ml, sterilized by membrane filtration (0.45 μm pore size, Millipore Filter Corp., Bedford, Mass.) and added to the above basal medium to give concentrations of myoglobin from 12.5 to 100 μg/ml.

Myoglobin becomes more effective for growth after autoclaving, although it does not effectively support growth at 12.5 μg/ml. The crude liver extract, at concentrations that support substantial growth, contained only 10 μg of hemeproteins per milliliter. However, autoclaved myoglobin at 100 μg/ml supports rapid reproduction and repeated subculture (ten serial transfers) of C. briggsae when it is substituted for heated liver extract.

A number of other proteins were tested in the same basal medium. Hemoglobin and cytochrome c were slightly more effective than myoglobin on a weight basis. Both compounds retained their activity after autoclaving. Non-hemeproteins such as β-lactoglobulin, bovine serum albumin, and soluble casein, each at a concentration of 100 μg/ml, were inactive (Table 1). The nutritional requirement for reproduction of Caenorhabditis briggsae under these conditions thus seems to be satisfied generally by hemeproteins, but not by protein in the absence of heme.

A question remains as to whether the active factor is the Fe-porphyrin moiety of the hemeprotein complex. A partial answer was obtained from experiments with hemin. Hemin chloride (Calbiochem, Los Angeles, Calif.) was dissolved in 0.1N potassium hydroxide, neutralized with hydrochloric acid to pH 7.8, diluted to 1 mg/ml, and sterilized by Millipore filtration in the same way as the myoglobin solution (Table 1).

81

At 8 μg/ml (equivalent to 200 μg of myoglobin on a molar basis), hemin was as effective as myoglobin at 100 μg/ml (Table 1); at 4 μg/ml hemin was approximately as effective as myoglobin at 50 μg/ml. At 2 μg of hemin per milliliter the generation time increased significantly and the total population was small. The 1 μg/ml concentration was ineffective. Therefore, on a molar basis, hemin is only about one-half as active as myoglobin although it is an effective substitute. One may speculate that the differences in effective doses could be related to problems of absorption; for example, precipitated versus soluble forms.

The results clearly demonstrate that hemin or heme compounds are required for reproduction of C. briggsae in a defined medium supplemented with sterols and autoclaved E. coli cells. Thus, one of the active components of liver extract (or of fractions derived from it) is most probably a hemeprotein. The meaning of the requirement for specialized structures reported by Sayre et al. (5) for their growth factor is unclear, especially since the material is replaceable by three contituents—a heme compound, a sterol, and a heat-stable component present in bacterial cells.

The requirement for heme may be related to the metabolism of heme-proteins in the parasitic nematode Ascaris lumbricoides. Smith and Lee (12) reported that the hemoglobin content of perienteric fluid increased when hemeproteins or porphyrins were added to the incubation medium; much hematin was incorporated into developing eggs of this organism.

References and Notes

1. M. Rothstein and W. L. Nicholas, in *Chemical Zoology*, M. Florkin and B. T. Scheer, Eds. (Academic Press, New York, 1969), vol. 3, p. 289.
2. F. W. Sayre, E. L. Hansen, E. A. Yarwood, *Exp. Parasitol.* 13, 98 (1963).
3. E. C. Dougherty, E. L. Hansen, W. L. Nicholas, J. A. Mollett, E. A. Yarwood, *Ann. N.Y. Acad. Sci.* 77, 176 (1959); W. L. Nicholas, E. C. Dougherty, E. L. Hansen, *ibid.*, p. 218.
4. F. W. Sayre, R. T. Lee, R. P. Sandman, G. Perez-Mendez, *Arch. Biochem. Biophys.* 118, 58 (1967).
5. F. W. Sayre, M. C. Fishler, G. K. Humphreys, M. E. Jayko, *Biochim. Biophys. Acta* 160, 63 (1968); F. W. Sayre, M. C. Fishler, M. E. Jayko, *ibid.*, p. 204.
6. W. F. Hieb and R. P. Sandman, *Fed. Proc.* 26, 797 (1967); R. Pertel, thesis, University of California (1967).
7. W. F. Hieb and M. Rothstein, *Science* 160, 778 (1968).
8. ——, *Arch. Biochem. Biophys.*, in press.
9. H. E. Snyder and J. C. Ayres, *J. Food Sci.* 26, 469 (1961).
10. E. J. Buecher, E. Hansen, E. A. Yarwood, *Proc. Soc. Exp. Biol. Med.* 121, 390 (1966).
11. Available as *C. briggsae* medium "75," Grand Island Biological Co., Grand Island, N.Y.
12. M. H. Smith and D. L. Lee, *Proc. Roy. Soc. London* 157B, 234 (1963).
13. We thank Dr. W. D. Brown for his advice and assistance with the electrophoresis and absorption spectra. Supported in part by PHS grants AM 12625 and AI 07145.

W. F. HIEB
MORTON ROTHSTEIN

Isolation from Liver of a Heat-Stable Requirement for Reproduction of a Free-Living Nematode[1]

Free-living nematodes grown in axenic culture require, in addition to a chemically defined medium, a heat-labile "growth factor" which is present in various tissues and bacteria (1, 2). A purified form of the factor derived from liver extract was reported by Sayre et al. (3) to be a discrete protein of high molecular weight which exists in multiple active forms of similar amino acid composition. Various treatments, such as freezing or warming, appear to be necessary to "activate" this material. Its biological activity has been related specifically to its structural conformation (4), and to the presence of sulfhydryl groups (5). Since this preparation contains small amounts of substances other than protein— e.g., lipids and nucleic acids—it seemed quite possible that in reality, more than one constituent was required by the nematodes. We have shown that sterols are essential for the continuous reproduction of the free-living nematode, Caenorhabditis briggsae, and indeed cholesterol was found to be present in the lipid moiety of the growth factor (6). Furthermore, the evidence below indicates that, in addition to sterols, the tissue supplement can be separated into two components: (1) a "growth-promoting factor" found in autoclaved bacterial cells, and (2) a heat-stable "reproduction factor" isolated from liver extract.

[1] This work was supported in part by United States Public Health Service Grant AI 07145.

C. briggsae grows rapidly in a buffered salt solution containing live Escherichia coli (ML 30 strain) and any of several sterols (6). If the E. coli cells are autoclaved before use (8 min at 124°C), neither growth nor reproduction of C. briggsae occurs in the sterol-buffered salt medium. However, the autoclaved cells do stimulate growth in defined medium (7) containing sterols and yeast extract (see below) as indicated by the larval assay method (1). Larvae reach the size of small adults in about 5 days, but they do not reproduce; no growth occurs in the medium without cells. This observation indicated that the requirements for growth and reproduction may be separable and that under the above conditions, autoclaved E. coli provides the growth component. A liver component was then isolated which contains the requirement for reproduction, but not growth.

Heated liver extract (HLE) was prepared as described by Sayre et al. (8) by heating lamb liver homogenate at 53° for 6 min. The HLE was applied to a column of Sephadex G-100 (Pharmacia, Uppsala, Sweden) and eluted with distilled water (Fig. 1). Several of the major fractions (Fig. 1) were sterilized by membrane filtration (0.45-μ pore size, Millipore Corp., Bedford, Mass.) and assayed by observing the rate of growth and reproduction of individual larvae inoculated into 0.25 ml of medium according to the procedure of Dougherty et al. (1). Samples of the respective liver fractions were added to an equal volume of medium containing all of the following ingredi-

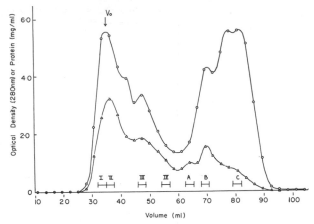

Fig 1. Gel filtration of heated liver extract on Sephadex G-100. A 3.0-ml vol of extract containing 33.8 mg protein/ml was applied to a 1.5 × 38-cm column and eluted with distilled water at a rate of approximately 6 ml/hr. The void volume (maximum exclusion volume) is designated as "V_0." Fractions were measured by absorbance at 280 nm (O—O), protein concentration by the method of Lowry *et al.* (10) (△—△). The fractions that were tested for their ability to support reproduction of *C. briggsae* are labeled I, II, etc. Spectra suggest that the red color of Fraction A is due to oxymyoglobin or oxyhemoglobin.

ents: one half strength defined medium (7) (available as *C. briggsae* medium "75," Grand Island Biological Co., Grand Island, New York), 50 μg/ml sterol mixture in Tween 80 as prepared previously (6), 2.5% yeast extract (Difco Laboratories) autoclaved for 8 min at 124°C, and approximately 0.3 mg/ml (dry weight) of autoclaved *E. coli* (6).

Of the peaks shown in Fig. 1, "Fraction A," which was red, gave rise to the most rapid reproduction of *C. briggsae* (about 5 days). In a control medium without autoclaved bacteria, Fraction A did not support reproduction. The results demonstrate clearly that both cells and Fraction A must be present for reproduction to occur. It was later found that the yeast extract was unnecessary in the medium.

The biological activity of "Fraction A" was further tested by a mass culture technique (9) (Fig. 2).[2] Large amounts of Fraction A were prepared by a chromatography procedure similar to the one described in Fig. 1. The heated liver extract was first concentrated approximately 6-fold by lyophilization and subsequent solution in water and applied to a 3.0 × 43-cm column of Sephadex G-100. The red fractions were combined, concentrated to the same volume as that of the initial HLE and added to the various media as

[2] Growth curves in Fig. 2 were obtained by Lan Sheng Huang.

indicated in Fig. 2. From the growth curves, it is apparent that continued and undiminished reproduction occurs in the medium containing autoclaved Fraction A and autoclaved *E. coli*. If either Fraction A or autoclaved cells are omitted, no reproduction occurs in the second subculture.

That Fraction A retains activity after autoclaving is extraordinary since all the previously active liver preparations were heat-labile (2, 3). A possible explanation lies in the successful separation of the growth factor into two essential components from different sources: autoclaved *E. coli* and autoclaved "Fraction A." It would, therefore, seem that the heat-labile nutrient present in live *E. coli* exists in a heat-stable form in the liver, and that the heat-labile component of liver extract exists in a heat-stable form in the bacteria.

From the above results, it is clear that the growth factor isolated by Sayre *et al.* (3) does not supply a single entity in the nutrition of *C. briggsae*.

The properties of the liver fraction (Fraction A) are different from those of the liver growth factor of Sayre *et al.* (3). The chromatographic properties of Fraction A suggest that it is much lower in molecular weight and it is clearly stable to autoclaving. Whether it is a protein or is protein-bound has not been established. In view of the heat stability of Fraction A, it is doubtful that the true growth factor for nematodes is a

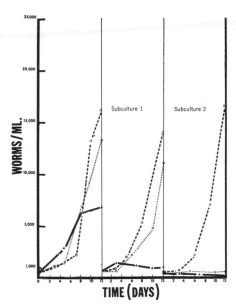

FIG. 2. Growth curves for *C. briggsae* in basal medium supplemented with the following: autoclaved "Fraction A" plus autoclaved *E. coli* (O—O); autoclaved *E. coli* (▲—▲); autoclaved "Fraction A" (□—□). Nematodes were grown in 50-ml Erlenmeyer flasks containing 5.0 ml of basal medium (similar to the totally defined medium, but with soy-peptone in place of amino acids) (9), 0.46 mg of cholesterol in Tween 80 (6), approximately 4 mg (dry weight) of *E. coli* cells and Fraction A equivalent to 0.7 ml of the original heated liver extract from which the fraction was isolated. All components of the medium were autoclaved together before inoculation with nematodes. Worm counts and sterility checks were made as previously reported (9).

protein of any special configuration as reported by Sayre *et al.* (4, 5).

REFERENCES

1. DOUGHERTY, E. C., HANSEN, E. L., NICHOLAS, W. L., MOLLET, J. A., AND YARWOOD, E. A., *Ann. N. Y. Acad. Sci.* **77**, 176 (1959).
2. NICHOLAS, W. L., DOUGHERTY, E. C., AND HANSEN, E. L., *Ann. N. Y Acad. Sci.* **77**, 218 (1959).
3. SAYRE, F. W., LEE, R. T., SANDMAN, R. P., AND PEREZ-MENDEZ, G., *Arch. Biochem. Biophys.* **118**, 58 (1967).
4. SAYRE, F. W., FISHLER, M. C., AND JAYKO, M.E., *Biochim. Biophys. Acta* **160**, 204 (1968).
5. SAYRE, F. W., FISHLER, M. C., HUMPHREYS, G. K., AND JAYKO, M. E., *Biochim. Biophys. Acta* **160**, 63 (1968).
6. HIEB, W. F., AND ROTHSTEIN, M., *Science* **160**, 778 (1969).
7. BUECHER, E. J., HANSEN, E. AND YARWOOD, E. A., *Proc. Soc. Exp. Biol. Med.* **121**, 390 (1966).
8. SAYRE, F. W., HANSEN, E. L., AND YARWOOD, E. A., *Exp. Parasitol.* **13**, 98 (1963).
9. TOMLINSON, G. A., AND ROTHSTEIN, M., *Biochim. Biophys. Acta* **63**, 465 (1962).
10. LOWRY, O. H., ROSEBROUGH, N. J., FARR, A. L., AND RANDALL, R. J., *J. Biol. Chem.* **193**, 265 (1951).

Heterotylenchus autumnalis

Hemocytic Reactions and Capsule Formation in the Host, *Musca domestica*

Anthony J. Nappi and John G. Stoffolano, Jr.

Encapsulation of nematode parasites in insects is a commonly observed phenomenon. Dying or dead nematodes have been seen enveloped by fat tissue, acellular fibrous material, connective tissue, and various types of homogeneous cytoplasmic masses invaded by tracheoles, malpighian tubules, and other elements. Typically, however, the reaction is characterized by the aggregation and fusion of host cells around the parasite and the intra- and/or extracellular deposition of pigment, generally believed to be melanin, on or near the surface of the parasite. Although most investigators consider the blood cells or hemocytes to be the principal encapsulating elements, there are relatively few studies of the actual formation of the capsule during infection, and the types and origin of capsule-forming cells. Reviews of the literature on the defense reactions of insects to nematode parasites are given by Salt (1963) and Poinar (1969).

The nematode *Heterotylenchus autumnalis* parasitizes the face fly, *Musca autumnalis*. The parasite is heteromorphic, having a gamogenetic stage alternating with a parthenogenetic stage. Mating of the nematodes and infection of host larvae occur in cattle manure, the breeding site of the fly.

Although there is evidence to suggest that *H. autumnalis* is distributed throughout the range of its host (Stoffolano, 1968), the parasite has not been reported from other dung-breeding Diptera in North America (Jones and Perdue, 1967). However, the parasite has been reported in two other *Musca* species in Czechoslovakia (Vilagiova, 1968).

Because of the possible role of the parasite as a biological control agent, studies were made on the parasite's host specificity to other dung-breeding flies. Attempts to establish the parasite in larvae of the house fly, *M. domestica*, were unsuccessful because of the defense reaction of the host. In order to determine the nature of the defense reaction of *M. domestica* to *H. autumnalis*, the hemocyte picture of host larvae was studied histologically during infection to show the relationship between hemocyte changes and capsule formation.

MATERIALS AND METHODS

Adult flies were kept in screened cages maintained in the insectary under continuous light at 27°C and 50% relative humidity. They were fed powdered milk, granulated sugar, and water. One-half gallon containers of fresh, rectally collected cow manure were placed into the cages as the oviposition medium. After oviposition the containers were removed from the cages and the larvae transferred to large enamel pans containing fresh manure with sand at one end for pupation.

Abdomens of 12-day-old infected, adult face flies were opened and nematode larvae spread over the surface of fresh manure in petri dishes. Nematodes were left in the dishes for 1 day to permit mating, and then 1-day-old house fly larvae were placed into the manure for 3–5 hr. The larvae were then removed, rinsed in saline and placed into enamel pans containing fresh manure. At 1, 3, and 5 days after infection, fly larvae were removed from the medium, rinsed in saline, and prepared for histological examination.

Controls consisted of larvae of comparable age not exposed to the parasites. Infected fly larvae were dissected at various times after infection. Whole mounts of the nematodes were examined under a phase-contrast microscope for hemocytic involvement and/or capsule formation.

Parasitized and nonparasitized larvae were fixed in Kahle's fixative (60°C). The anterior portion of the larva, just behind the mouth hooks, was removed to permit better penetration of the fixative. Larvae were dehydrated in ethyl alcohol, cleared in benzene, and infiltrated in Tissuemat (56°C, mp). Longitudinal sections (8 μ) were made and stained in Delafield's hematoxylin and eosin Y.

Differential hemocyte counts were made from serial sections 1 and 3 days after infection. A minimum of 20 sections of each larva was examined. The sections in which the counts were made were selected at random, and whenever possible, a minimum of 100 cells was counted. Prohemocytes were included with the plasmatocytes for the differential hemocyte counts because transitional forms made it difficult to distinguish between large prohemocytes and small plasmatocytes.

After the differential counts were taken, the slides were reexamined for a comparative study of the mitotic indices. Approximately 1000 hemocytes from each larva were examined, and the number of cells with mitotic figures was recorded. Generally, 10 sections (100 cells/section) of each larva were studied.

RESULTS

General Observations

House fly larvae, pupae, and adults dissected at various intervals after parasitization showed that only the infective, gamogenetic stage of the parasite was encapsulated. In no case were any other stages, including eggs, encapsulated.

An early indication of a defense reaction

was the appearance of a melanic deposit or "cap" covering the mouth of the nematode. Examinations of nematodes during the course of infection showed pigment deposited on other areas of the cuticle. In some infected larvae nematodes were found near the caudal spiracles partially enveloped by a sticky, opaque white matrix.

Histological Observations

In longitudinal sections of 2-, 4-, and 6-day-old nonparasitized larvae the hemocytes were aggregated in the last two abdominal segments, usually between the muscle of the body wall and the lateral and posterodorsal epidermis. The hemocytes were free or in small groups, but were not surrounded by a membrane. During larval development these posterior hemocytic masses increased in size. Prohemocytes, plasmatocytes, and oenocytoids were seen within these masses in various stages of differentiation, but only the prohemocytes and plasmatocytes were noted in division. Except for the presence of a few hemocytes scattered throughout the body, the only evidence to suggest that the hemocytes circulated was found in 6-day-old larvae. In these, a relatively large mass of hemocytes was seen just below the proventriculus. Although total hemocyte counts were not made, it was obvious that the number of hemocytes increased during larval development.

Histological examinations of house fly larvae 1 day after infection showed that in regions where the parasites penetrated the host the cuticle was melanized. In some hosts, the epidermal cells below the areas of cuticular melanization were enlarged and contained a brown to black pigment. Nematodes were usually found between the epidermis and underlying muscles, and near the alimentary canal in the posterior abdominal segments. However, none of the parasites 1 day after infection was encapsulated or melanized. Evidence of a hemocytic reaction at this time was the appearance of small, irregularly distributed aggregations of hemocytes throughout the abdomen of the host. In addition, several small syncytia or multinucleate masses, each composed of 2–4 fused hemocytes, were seen in the posterior hemocytic areas (Figs. 1, 2). The hemocytes in the developing syncytia were mostly oenocytoids and a few plasmatocytes.

The first evidence of a successful defense reaction was noted in house fly larvae 3 days after infection. In these hosts the nematodes were heavily melanized and encapsulated (Figs. 6–8). In all cases the capsule material was superficial to the pigment which appeared to have been deposited first directly on the cuticle of the parasite. Examination of the capsules showed the enveloping material to be a homogeneous mass of cellular origin, and identical to the hemocyte syncytia seen earlier in larvae 1 day after infection. In parasitized larvae 3 and 5 days after infection the syncytia were larger (20–30 fused cells) and more numerous than those found 1 day after infection, and many had encapsulated parasites (Figs. 3–5). The formation of syncytia was not limited to the posterior hemocytic areas, or to areas immediately surrounding parasites, for small syncytia were seen elsewhere in the body of the host, and not near parasites.

In some cases the cellular capsules were invaded by connective tissue strands, tracheae, tracheoles, muscle epithelium, and fat body cells (Figs. 6, 7). The tracheal epithelium and fat body cells frequently showed signs of disintegration and pigmentation, even when they were not part of the capsule material. In one parasitized larva encapsulated nematodes were found in the dorsal blood vessel (Fig. 8).

The total number of hemocytes in parasitized larvae 3 and 5 days after infection appeared to be greater than in the controls, for large masses of hemocytes were found irregularly distributed in the hemolymph of infected larvae (Figs. 9–11). In addition,

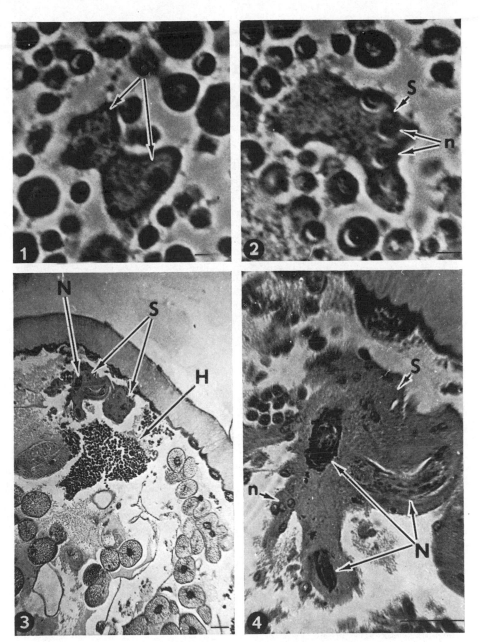

Figs. 1–4. Photomicrographs of hemocyte syncytia (S) in the posterior hemocytic areas of parasitized larvae of *M. domestica*. Figs. 1 and 2. One day after infection, showing the fusion of oenocytoids (O), remaining nuclei (n), and the early development of a syncytium. Bar = 10 μ. Figs. 3 and 4. Three days after infection. Fig. 3. Two syncytia in the posterior hemocytic area containing encapsulated nematodes (N) and showing aggregation of hemocytes (H). Bar = 0.5 mm. Fig. 4. Highly magnified section of syncytium seen in Fig. 3, showing remaining nuclei (n) and encapsulated nematodes (N). Bar = 0.5 mm.

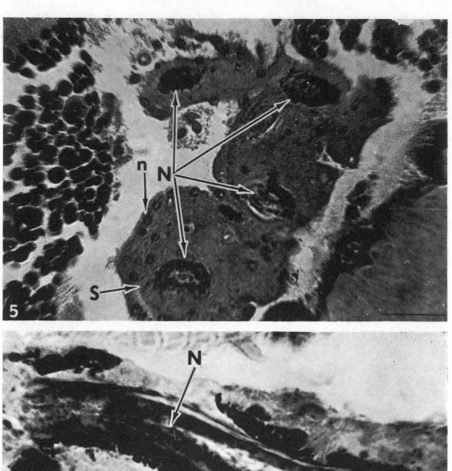

Figs. 5 and 6. Photomicrograph of a large syncytium (S) showing encapsulated nematodes (N), and nuclei (n) of hemocytes. Bar = 0.5 mm. Fig. 6. Longitudinal section of *H. autumnalis* (N) surrounded by a pigmented layer or "sheath capsule" (C) and a cellular capsule (= syncytium)(S). Bar = 10 μ.

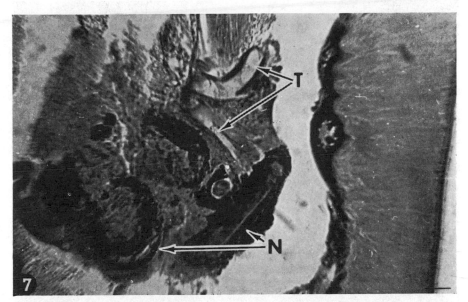

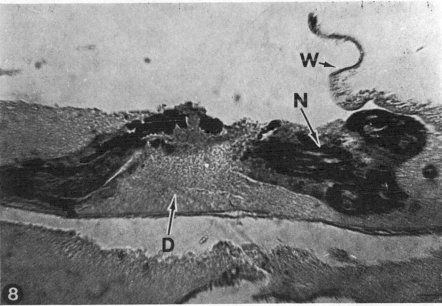

Figs. 7 and 8. Photomicrograph showing encapsulated nematodes (N) with tracheae (T) incorporated into the cellular capsule. Fig. 8. *H. autumnalis* (N) melanized and encapsulated within the dorsal blood vessel (D) of *M. domestica*. Note the ruptured wall (W) of the blood vessel. Bars = 10 μ.

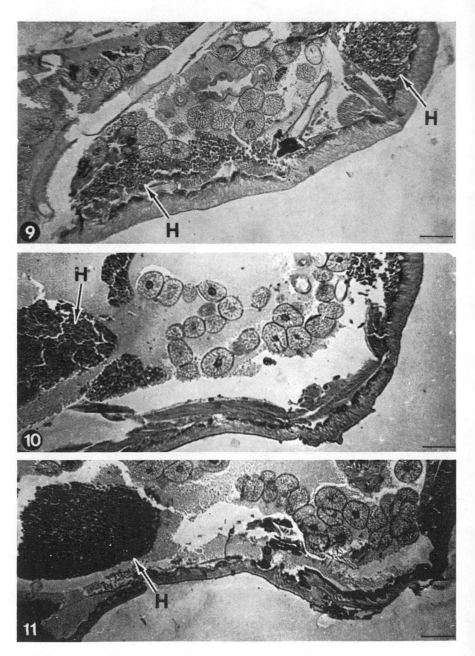

FIGS. 9–11. Photomicrographs of parasitized larvae of *M. domestica* 3 days after infection showing large masses of hemocytes (H) irregularly distributed throughout the hemocoel. Bars = 0.1 mm.

the posterior hemocytic masses in parasitized larvae were larger than those in nonparasitized individuals. Also important was the presence of small aggregations of prohemocytes and plasmatocytes near the ventral epidermis of the second from the last abdominal segment of parasitized larvae only. The epidermal cells in this area were vacuolated and showed signs of proliferation and disintegration.

Differential Hemocyte Counts

In addition to the apparent increase in the total number of cells in parasitized larvae, it was considered important to determine what changes, if any, occurred in the hemocyte complex of these larvae during the early stages of infection, i.e., during the formation of syncytia and the melanization and encapsulation of the parasites. Differential hemocyte counts were made of control and parasitized larvae 1 and 3 days after infection. Results of these counts

TABLE I

Differential Hemocyte Counts from Parasitized and Nonparasitized Larvae of Musca domestica

Specimen number	Nonparasitized[a]			Specimen number	Parasitized[a]		
	No. cells	% p–pl	%O		No. cells	% p–pl	%O
1 Day after infection							
1	4578	89.5	10.4	5	4588	96.0	3.9
2	3404	90.5	9.4	6	4402	96.8	3.1
3	3410	92.6	7.3	7	2798	97.9	2.0
4	3666	92.3	7.7	8	3253	93.3	6.6
Mean		91.1	9.0			96.0	3.9
3 Days after infection							
9	4912	89.7	10.2	13	2978	97.2	2.7
10	4522	89.6	10.4	14	5205	97.1	2.8
11	4538	87.8	12.1	15	4927	96.9	3.0
12	4545	88.5	11.4	—	—	—	—
Mean		88.9	11.0			97.0	2.9

[a] p–pl = prohemocytes plus plasmatocytes; O = oenocytoids.

TABLE II

Mitotic Indices Among Hemocytes from Parasitized and Nonparasitized Larvae of Musca domestica

Specimen number	Nonparasitized		Specimen number	Parasitized	
	No. cells	% Mitosis		No. cells	% Mitosis
1 Day after infection					
1	1095	0.09	5	1033	0.29
2	1004	0.19	6	1054	0.09
3	1009	0.39	7	991	0.30
4	1018	0.39	8	989	0.50
Mean mitotic rate		0.26			0.29
3 Days after infection					
9	1016	0.49	13	1056	0.56
10	1000	0.40	14	1000	0.30
11	1020	0.39	15	1016	0.68
12	995	0.20	—	—	—
Mean mitotic rate		0.37			0.52

showed the percentages of oenocytoids in all parasitized larvae of both age groups were less than in the control larvae (Table I). Oenocytoids averaged 3.9% in larvae 1 day after infection and 2.9% in larvae 3 days after infection, compared with 9.0% and 11.0%, respectively, in the controls.

A comparative study of the percentages of dividing hemocytes in control and parasitized larvae showed only slight differences (Table II). The mitotic index averaged 0.29% in larvae 1 day after infection, and 0.52% in larvae 3 days after infection, compared with 0.26% and 0.37%, respectively, in the controls. The average mitotic index of all parasitized larvae was 0.39% (28/7139), while in the controls it was 0.31% (26/8157).

DISCUSSION

The capsule formed by *M. domestica* larvae around the infective stage of the nematode parasite *H. autumnalis* is of hemocytic origin. Histological observations of parasitized larvae at various stages of infection suggest that the initial host reaction includes the aggregation, and fusion of hemocytes to form a multinucleate mass that adheres to the cuticle of the parasite. Concurrently, melanin is deposited on the parasite and within the substance of the developing capsule forming a thin, pigmented layer around the parasite. Hemocytes continue to aggregate and fuse around the melanized nematode forming a thick, homogeneous mass or syncytium containing hemocyte nuclei. The completed capsule consists essentially of two parts; an inner pigmented layer appressed to the cuticle of the nematode, and an outer syncytial mass of host hemocytes (cellular capsule). The pigmented inner layer is similar to the "melanized sheath" formed by *Aedes aegypti* around rhabditoid nematodes (Bronskill, 1962), and to the "sheath capsule" described by Salt (1963).

In some cases connective tissue strands, tracheae, tracheoles, muscle epithelium, and fat body cells are incorporated into the hemocytic capsules. Although these structures have been implicated in the encapsulation reactions of numerous insects infected by helminths (c.f. Salt, 1963; Poinar, 1969), these elements are not common or essential components of the capsule formed by *M. domestica* larvae around *H. autumnalis*.

Differential hemocyte counts taken before and during capsule formation show that the percentages of oenocytoids in all parasitized larvae are less than in the controls. The decrease in the percentage of oenocytoids, at a time when these cells fuse to form syncytia, and pigment is deposited on the parasite, suggest that these cells are responsible for the formation of the pigmented layer that characterizes the initial host reaction, and for the aggregation and fusion of other hemocytes forming the enveloping cellular capsule.

Although total hemocyte counts were not made, it was evident that the number of hemocytes in parasitized larvae was greater than in the controls. Unfortunately, the manner in which the blood cells increased in number in parasitized larvae is not known. Although the posterior hemocytic areas are considered to be hemocytopoietic foci (Arvy, 1954), comparative studies of the percentages of dividing hemocytes in control and in 1- and 3-day infected larvae show little difference in the mitotic indices (Table II). Since virtually all hemocyte activity relative to melanization and capsule formation occurred during this period, it seems unlikely that the slight increases in the percentages of dividing cells (0.03–0.15%) in parasitized larvae could alone account for the disparity noted in the total number of cells.

Examination of parasitized larvae for other possible hemocytopoietic areas showed a small mass of hemocytes near the ventral epidermis of the second from the last abdominal segment. The epidermal cells in this region were vacuolated and showed signs of disintegration and proliferation. However, it could not be determined if the hemocytes had originated from the epidermal cells, or if they were instead responsible, by phagocytic activity, for the degenerative appearance of the epidermis.

Although most insect hematologists consider hemocytes to be of mesodermal origin, Jones (1970) pointed out that in many instances tissues which may be potential hemocytopoietic foci have not been studied experimentally. It may be that during conditions of stress, i.e., infection, certain insect tissues become formative and respond to infectious agents, microbial or macrobial. In this connection it is of interest to note that Couturier (1963) reported that the capsules formed by *Melolontha melolontha* larvae around mermithid nematodes were produced from the proliferation and multiplication of epidermal cells which were irritated by the penetrating nematode. Also, Bronskill (1962) found epidermal cells of adult *Aedes aegypti* growing around encapsulated rhabditoid nematodes, isolating the capsules from the host's hemocoel.

In *M. domestica* capsule formation about spiruroid nematodes of the genus *Habronema* has been studied by several workers (Hill, 1918; Roubaud and Descazeaux, 1921, 1922; Poinar, 1969). In house fly pupae and adults Poinar (1969) found the infective stage of *H. muscae* embedded in a homogeneous mass of granular cytoplasm that was pigmented and interlaced with tracheoles, but he could not determine if hemocytes contributed to the formation of the capsule. Roubaud and Descazeaux (1921) reported that when *H. megastoma* came in contact with the malpighian tubules of host larvae, the epithelium proliferated around the parasite to form a "thylacie" (= "pseudocyst"). Presumably, phagocytic hemocytes later invaded the thylacie and brought about the disintegration of the cellular elements of the epithelium. Some thylacies remained attached to the tubules, while others were free in the hemocel of the host. Roubaud and Descazeaux suggested that the thylacies, produced by this tumorous reaction of the epithelium, protected the parasites from histolysis during metamorphosis of the host. The nematodes developed within the thylacies and later emerged to infect tissues of the adult fly. However, a careful study of the authors' diagrams show that, except for their mode of formation, these multinucleate masses are identical to the hemocyte capsules reported in this study.

The possibility that thylacies are capsules and not cysts and, thus, represent a defense reaction of the host should be examined. The fact that the *Habronema* successfully emerged from these enveloping structures could be attributed to an unsuccessful host reaction, or to the ability of the parasite to resist, to varying degrees, the reactions of the host.

REFERENCES

ARVY, L. 1954. Données sur la leucopoièse chez *Musca domestica* L. *Proceedings of the Royal Entomological Society of London Series A* **29**, 39–41.

BRONSKILL, J. F. 1962. Encapsulation of rhabditoid nematodes in mosquitoes. *Canadian Journal of Zoology* **40**, 1269–1275.

COUTURIER, A. 1963. Recherches sur des Mermithidae Nématodes du Hanneton commun (*Melolontha melolontha* L. Coleopt. Scarab.). *Annales des Epiphyties* **14**, 203–267.

HILL, G. F. 1918. Relationship of insects to parasitic diseases in stock. *Proceedings of the Royal Society of Victoria* **31**, 11–107.

JONES, C. M. AND J. M. PERDUE. 1967. *Heterotylenchus autumnalis*, a parasite of the face fly. *Journal of Economic Entomology* **60**, 1393–1395.

JONES, J. C. 1970. Regulation of Hematopoiesis (Ed.) A. S. Gordon. Vol. 1, 7–65. Appleton-Century-Crofts, New York.

POINAR, G. O., JR. 1969. Arthropod immunity to worms. *In* "Immunity to Parasitic Animals" (G. J. Jackson, R. Herman, and I. Singer, eds.), Vol. 1, pp. 173–210. Appleton-Century-Crofts, New York.

ROUBAUD, E. AND J. DESCAZEAUX. 1921. Contribution à l'histoire de la mouche domestique comme agent vecteur des habronémoses d'equidés. Cycle évolutif et parasitisme de l'*Habronema megastoma* (Rudolphi, 1819) chez la mouche. *Bulletin de la Societe de Pathologie Exotique* **14**, 471–506.

ROUBAUD, E. AND J. DESCAZEAUX. 1922. Deuxieme contribution à l'etude des mouches, dans leurs rapports avec l'évolution des habronemes d'equidés. *Bulletin de la Societe de Pathologie Exotique* **15**, 978–1001.

SALT, G. 1963. The defense reactions of insects to metazoan parasites. *Parasitology* **53**, 527–642.

STOFFOLANO, J. G., JR. 1968. Distribution of the nematode *Heterotylenchus autumnalis*, a parasite of the face fly, in New England with notes on its origin. *Journal of Economic Entomology* **62**, 792–795.

VILAGIOVA, I. 1968. *Heterotylenchus autumnalis* Nickle (1967)—a parasite of pasture flies. *Biologia Bratislava* **23**, 397–400.

Ultrastructure of Nematode Tissues

CUTICULAR FINE STRUCTURE AND MOLTING OF *NEOAPLECTANA GLASERI* (NEMATODA), AFTER PROLONGED CONTACT WITH RAT PERITONEAL EXUDATE*

George J. Jackson and Phyllis C. Bradbury

The cuticle that covers and supports nematodes is remarkably resistant to destruction by chemical and physical means. In autolysis or in attack by the digestive processes of other organisms, whether cellular or extracellular, it is most often the cuticle that persists structurally after a nematode's soft tissues have begun to change. However, when this work was begun, possible changes in cuticular fine structure caused by adhering, presumably phagocytic cells had not been looked for with electron microscopy.

Neoaplectana glaseri Steiner, 1929, not naturally a parasite of mammals, is a convenient nematode for such studies. Large numbers of worms, uniform as to stage, are available from species isolation cultures (Glaser, 1940); certain effects of mammalian temperature, antibody, and cells on the worms have been described and distinguished in culture (Jackson, 1961, 1962, 1966).

MATERIALS AND METHODS

The natural and laboratory histories of *Neoaplectana glaseri* and methods for culturing this nematode in species isolation have been detailed (Stoll, 1959, 1961; Jackson, 1969). To obtain a maximal exudate, 5,000 to 6,000 living third-stage nematode larvae contained in 8 ml of Abbott "Isotonic Sodium Chloride Solution" (900 mg NaCl/100 ml H$_2$O) with 0.9% glycogen were injected into the peritoneal cavity of 100-g female Sprague–Dawley rats. On one occasion 25-g rats were used. Glycogen increased the number of cells in the peritoneal exudate compared to injections of living worms suspended in saline. Injections of dead worms in saline stimulate slight exudates of short duration but seldom can initially dead worms be recovered from the rat after 24 hr.

From previous unpublished results it was known that, in contrast to *Turbatrix aceti* cultures, many *N. glaseri* survived at least 24 hr in the rat peritoneal cavity. Consequently, the "host" population was sampled at daily intervals for 7 days after the initial injection. For these daily samples, 3 anesthetized rats were each injected with 6 ml of the Abbott saline, the rib cage area was massaged, and, after a rat had been killed with ether, the saline was withdrawn with a Pasteur S-C pipette (Stoll, 1961) through a small incision into the peritoneal cavity.

Worms in the withdrawn samples were allowed to settle without centrifugation in pointed tubes under sterile conditions. Some of the worms from pooled samples were examined by light microscopy; some were transferred to an optimal liquid medium (Stoll, 1961) for culturing; others were fixed for electron microscopy in cold (ca. 0 C) 3% glutaraldehyde, 0.1 M cacodylate. In this mixture the worms were cut in half with a stainless steel razor blade or processed whole until the final sectioning. After 1 hr, both whole and halved worms were washed for at least an hour with several changes of cold 0.1 M cacodylate, 4% sucrose. For postfixation the worms were transferred for 1 hr to cold 2% osmium in 0.2 M sucrose buffered at pH 7.3 with 0.1 M cacodylate before dehydration and embedding in the following series: 70%, 95%, 3 × 100% ethanol; 3 × propylene oxide (10 min for all preceding steps); 1/1, 1/2 propylene oxide and Epon resin (with DMP-30 catalyst to 1.5%), (45 min each); repeated changes in fresh Epon–DMP over several days at room temperature to aid penetration of the resin. The worms were then oriented for convenient sectioning in Beem capsule covers and cured at 52 C for 3 to 5 days. The

* Supported in part by the U. S. Public Health Service through grants AI-04842, K3-AI-9522, and TO1-AI-00192 from the NIAID.

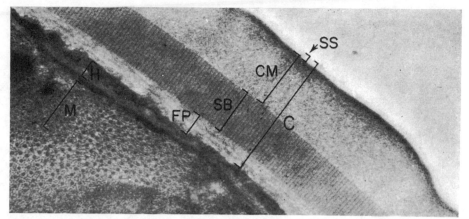

FIGURE 1. *Neoaplectana glaseri* third-stage larva from culture, showing section of cuticle (C) with its surface sheet (SS), cortex mat (CM), striation band (SB), and fibrous pad (FP). Proximally are hypodermis (H) and musculature (M). × 37,000.

embedded worms were sectioned with a diamond knife on a Porter–Blum MTI ultramicrotome and examined in an RCA (EMU-3F) electron microscope. Some sections, after the usual mounting on grids, were stained with 10% phosphotungstic acid for 2 hr then washed with methanol; others were stained with uranyl acetate and lead citrate.

For light microscopy of the cellular exudate some saline samples flushed from the peritoneal cavity of injected rats were centrifuged, the supernatant discarded, and the sediment mixed with a drop of 2% bovine serum albumin, spread on glass slides, air dried, and fixed in methanol, then stained with pH 6.8 buffered Giemsa.

RESULTS

For 4 consecutive days living and dead *Neoaplectana glaseri* were flushed in decreasing numbers from the peritoneal cavity of rats weighing about 100 g on the day of injection. On day 4, only about 1% of injected worms could be recovered. No whole, recognizable worms were recovered thereafter except for day 5 on the one occasion when 25-g rats had been used. The majority of the worms were still third-stage larvae although, unlike the inoculum, most were now exsheathed. Rarely had a recovered worm developed beyond third stage; no mature adults were seen.

Third-stage *N. glaseri* flushed from the rat peritoneal cavity during the first 3 days after injection and placed in culture developed into adults irregularly but did not reproduce. Third-stage larvae recovered live on day 4 survived for about 2 weeks in culture but did not develop to a more mature stage.

Many host cells were so firmly attached to worms on removal from the rat that washing in saline did not detach them, but after several days in culture tubes they generally dropped off. A few recovered worms were additionally enmeshed in strands of fibrous material from which they did not free themselves readily. Light microscopy of stained peritoneal exudate from rats injected with the saline suspension of worms and glycogen showed many and varied mononuclear and polymorphonuclear leukocytes during the 1st week after injection. The total number of retrieved cells decreased after the 3rd day although one morphological type increased at this time: a small mononuclear cell with extremely basophilic staining properties and little cytoplasm. According to light microscopy of worms on these Giemsa-stained slides, the adhering cells were of no one type; this was also found in electron microscopy.

Electron microscopy shows that the cuticle of third-stage *N. glaseri* is a multilayered structure similar to cuticles of other nematodes. The number of layers and sublayers varies even in uniformly prepared material. Avoiding, therefore, as much as possible the assumptions involved in distinguishing between real "layers"

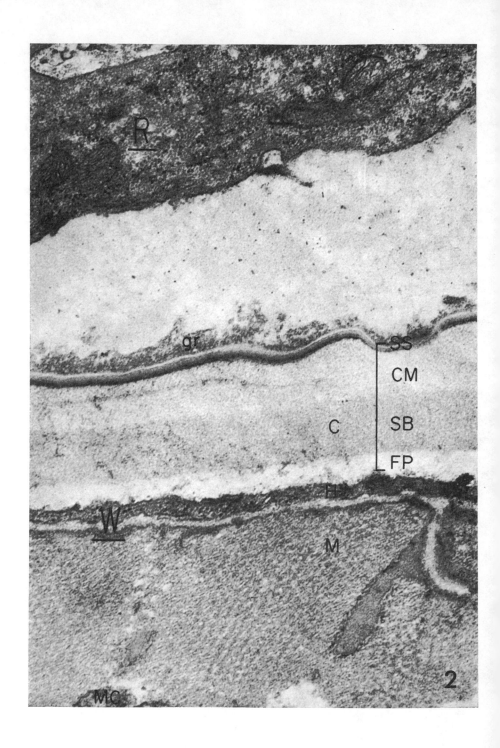

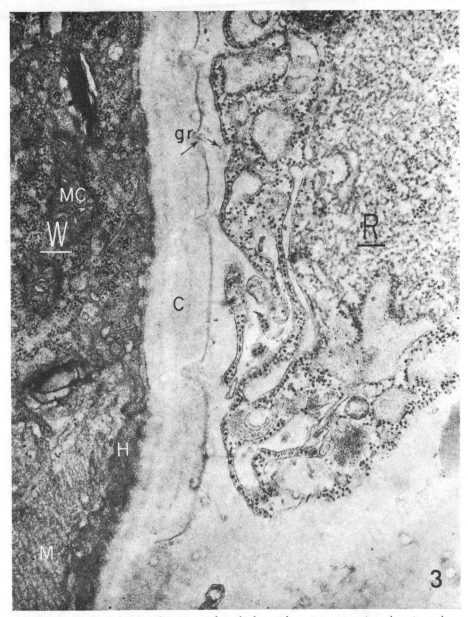

Figures 2, 3. *Neoaplectana glaseri* injected as third-stage larva into rat peritoneal cavity and recovered after 3 days. Showing section of worm (W) and adhering rat cell (R), probably monomorphonuclear leukocyte. Granular material (gr) on surface of worm cuticle or between cuticle and rat cell. In these preparations layers of the cuticle not prominent. Worm tissues inside cuticle, including musculature (M), hypodermis (H), and mitochondria(MC) appear normal. Figure 2 × 45,000; Figure 3 × 21,000.

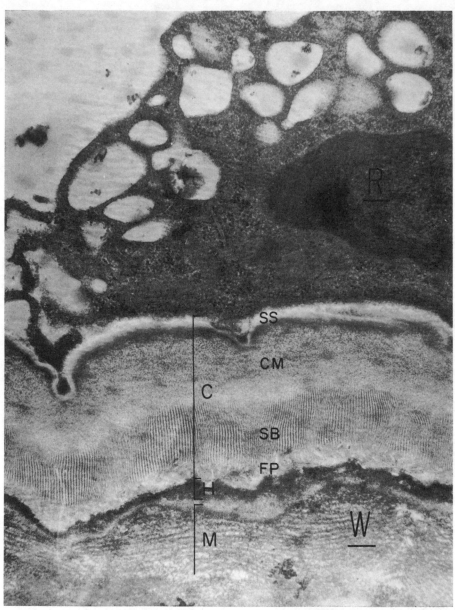

FIGURE 4. *Neoaplectana glaseri* injected as third-stage larva into rat peritoneal cavity and recovered after 4 days. Showing section of worm (W) and adhering rat cell (R) with no space between. Cells probably monomorphonuclear leukocytes ("macrophages"). Surface sheet (SS) of worm cuticle swollen and normally regular pattern of striation band (SB) distrupted. Below cuticle (C), hypodermis (H) and musculature (M) appear intact. × 45,000.

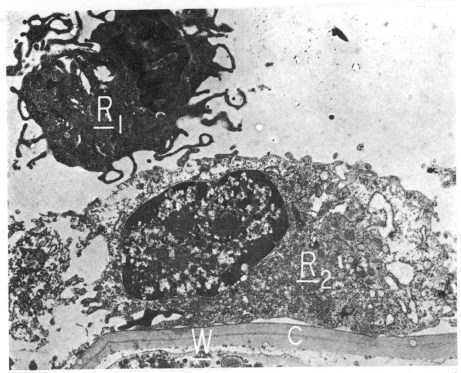

FIGURE 5. *Neoaplectana glaseri* injected as third-stage larva into rat peritoneal cavity and recovered after 3 days. Showing two different rat cell types (*R1*, *R2*); either type may adhere directly to the cuticle (C) of the worm (W), or be attached to a directly adhering cell. × 11,000.

and mere "sublayers," we frequently see (Fig. 1) four longitudinal regions in sections of third-stage *N. glaseri* cuticle: a thin, surface sheet (SS), a thick cortex mat (CM), a striation band (SB), and proximally a fibrous pad (FP). The presence of other layers or the absence of those already mentioned will be noted where appropriate.

Host cells adhered to all *N. glaseri* flushed from the peritoneal cavity of rats on days 1 to 5 following injection. In some sections there was no space between the cell and the cuticle for much of their common border (Fig. 4). In other sections the space between rat cell and worm was probably "real" because a cut had been made at the rounded edges of cells most of whose adhering surface was flattened against a worm (Fig. 3); some preparations showing a space were probably due to

preparative shrinkage (Fig. 2). A granular substance (gr) was seen on the cuticle of recovered worms where no cells had become attached (or had dropped off) or where there was spatial separation of the rat cell and the worm (Figs. 2, 3). Identification of cell types in sections that show more worm than rat cell is not always possible but, as in the light microscopy of peritoneal exudates stained with Giemsa, the electron microscopy also suggests that more than one host cell type adhered to the surface of worms (Fig. 5). In general the mitochondria and ribosomes of adhering cells were morphologically intact.

Worms removed during the first 3 days after injection did not, despite cell coats, show an altered cuticle. Although not all four layers described in uninjected worms could be seen or seen easily in some of these sections this cannot, as already mentioned, be ascribed to the host's influence. Below the cuticle no degenerative changes were apparent in such worm tissues and structures as hypodermis (H), musculature (M), and mitochondria (MC).

Two cuticular changes were uniformly seen in the last worms, on days 4 and 5, recovered from injections into the rat peritoneal cavity. The surface sheet was swollen and the striation pattern was disrupted (Fig. 4). Like worms collected on preceding days, these worms were almost totally coated with host cells.

DISCUSSION

That electron microscopy showed no morphological changes in the cuticle of *Neoaplectana glaseri* heavily coated with host cells on recovery from the rat peritoneal cavity during 3 days following their injection can be correlated with the worms' persistent if limited ability to develop when replaced in culture. Crandall and Arean (1967), in electron microscopy of *Ascaris suum* larvae after 4 hr in the peritoneal cavity of immunized mice, also found adhering cells but no cuticular alterations. Soulsby (1963, 1966) and Morseth and Soulsby (1969) described host cell adhesion to third-stage *A. suum* as immunologically specific with "transformed" lymphocyte cells but not with granulocytes. In vitro findings in this laboratory (Jackson, 1966) suggested that host cells which might otherwise not stick to worms may be passively incorporated into immune precipitates caused by the reaction of host antibodies with worm secretions and excretions at the worm's surface.

No attempt was made to preimmunize rats in the present work and the peritoneal exudate is basically a "foreign body response" enhanced by glycogen. In addition, however, the onset of acquired immunity and even the production of antibodies by day 4 is not impossible. Worms replaced in culture after 4 days in the rat peritoneal cavity survived but did not develop to a more mature stage. This effect cannot be ascribed solely to unfavorable temperature. *N. glaseri* cultured for 1 week as 37 C mature irregularly (Jackson, 1961). However, oxygen scarcity or conditions in the rat peritoneal cavity other than those involved in acquired immunity might be damaging to the worms.

It is, however, most likely that rat cell action is responsible for the cuticular changes in nematodes after 4 or 5 days in the peritoneal cavity. Swelling of the surface sheet and disruption of the striation pattern are not the cuticular changes involved in normal molting, and do correlate with larval worms' failure to molt again when placed in an optimal culture environment. Whether the "degenerative" changes are caused by the enzymatic activity of the rat cells, or merely by their interference with "cuticular transport," cannot be decided without further work.

The nature of the granular material on the surface of all worms recovered from the rat peritoneal cavity was not determined. Probably it is of host origin but whether it is a blood constituent that helps bind the rat's cells to the worm's cuticle and/or has enzymatic activity for cuticular components remains to be seen.

It is also uncertain why the number of nematode cuticular layers that are easily seen with electron microscopy varies in uniformly prepared material. Are the differences due to "transitory" physiological states of the worm or worm area, or to a more constant "regional" morphology, or to an artifact?

LITERATURE CITED

CRANDALL, C. A., AND V. M. AREAN. 1967. Electron microscope observations on the cuticle and submicroscopic binding of antibody in *Ascaris suum* larvae. J. Parasit. **53:** 105–109.

GLASER, R. W. 1940. The bacteria-free culture of a nematode parasite. Proc. Soc. Exp. Biol. Med. **43:** 512–514.

JACKSON, G. J. 1961. The parasitic nematode, *Neoaplectana glaseri*, in axenic culture. I. Effects of antibodies and anthelminthics. Exp. Parasit. **11:** 241–247.

———. 1962. The parasitic nematode, *Neoaplectana glaseri*, in axenic culture. II. Initial results with defined media. Exp. Parasit. **12:** 25–32.

———. 1966. Serological and cultivational comparisons of *Neoaplectana* species, nematodes of insects. Proc. 1st Internatl. Cong. Parasit. (A. Corradetti, ed.; Pergamon Press, Oxford and Tamburini Editore, Milano) **1:** 578–579.

———. 1969. Nutritional control of nematode development. Adv. Exp. Med. Biol. **3:** 333–341.

MORSETH, D. J., AND E. J. L. SOULSBY. 1969. Fine structure of leukocytes adhering to the cuticle of *Ascaris suum* larvae. I. Pyronino-phils. J. Parasit. **55**: 22–31.

SOULSBY, E. J. L. 1963. The nature and function of the functional antigens in helminth infections. Ann. N. Y. Acad. Sci. **113** (Art. 1): 492–509.

———. 1966. The mechanisms of immunity to gastrointestinal nematodes. *In* E. J. L. Soulsby (ed.), Biology of Parasites. Academic Press, New York, p. 255–276.

STOLL, N. R. 1959. Conditions favoring the axenic culture of *Neoaplectana glaseri*, a nematode parasite of certain insect grubs. Ann. N. Y. Acad. Sci. **77** (Art. 2): 126–136.

———. 1961. Favored RLE for axenic culture of *Neoaplectana glaseri*. J. Helm., R. T. Leiper Suppl.: 169–174.

Microvilli on the Outside of a Nematode

The females of insect parasitic nematodes of the order Tylenchida are bizarre in form and live in the haemocoels of their hosts[1-3]. After the infective larva enters the haemocoel it usually moults, grows rapidly and becomes sexually mature. Its feeding apparatus quickly degenerates, the mouth and anus often disappear and if the intestine persists it loses its connexion with the oesophagus[4] when food must presumably be absorbed through the body wall. In keeping with this function the external surface of one of these parasites was found to be quite unlike that of nematodes so far described.

The nematode studied, *Bradynema* sp., was an undescribed parasite of the mushroom pest *Megaselia halterata* (Diptera, Phoridae)[5,6]. The adult female reproduces inside the haemocoel of the fly and second stage larvae leave the host during oviposition or defaecation. The larvae rapidly moult in the mushroom compost and develop into adult males and infective females within 48 h at 25° C. After copulation the female enters a second or third instar maggot where it grows and lays eggs in about 7 days.

Flies were dissected in entomological saline and the worms transferred to either 2·5 per cent glutaraldehyde in cacodylate buffer or osmium tetroxide buffered according to Millonig or Zetterquist. The adult female parasites were 1 mm long and each was cut into three or four pieces and left in the fixative for 2 to 24 h. Those fixed in glutaraldehyde were rinsed in cacodylate buffer and post fixed in either Millonig's or Zetterquist's osmium tetroxide for a further 1 to 4 h. They were rinsed in the appropriate buffer, orientated in agar[7], dehydrated in alcohols and slowly transferred to pure epoxy resin through six mixtures of increasing concentration. They were left overnight at 25° C before being transferred to fresh resin for polymerization at 60° C for two days. Pale gold sections were cut with glass knives on a Reichert ultramicrotome, mounted on uncoated copper grids and examined with a Jem 7 electron microscope at 80 kV.

Fig. 1 shows microvilli on the outside of the worm which cover the entire surface. The density, shape and size of microvilli differ according to their position and the age of the worm; some are simple finger-like projections, others have enlarged tips and several have branches which may

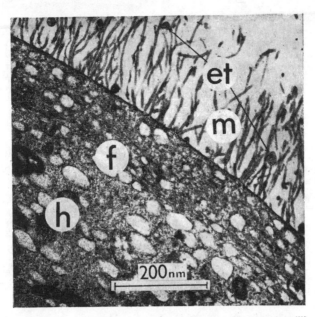

Fig. 1. Transverse section of the body wall surface showing microvilli (m) on the outside, some with enlarged tips (et) and supported by fibres(f) below in the hypodermis (h). The nematode was fixed for 3 h in Zetterquist's osmium tetroxide and stained with uranyl acetate and lead citrate.

anastomose with adjacent microvilli. Scattered fibres occur below the surface together with endoplasmic reticulum, lipid droplets, membranous whorls presumably of phospholipid, mitochondria, ribosomes and occasional nuclei, which suggest that the cuticle is absent and this region is the hypodermis. Two to eight cells in each complete transverse section contain what appear to be sparse myofilaments and probably represent degenerate muscle cells.

The body wall of nematodes is composed of three main regions: cuticle, hypodermis and muscle[8-10]. The cuticle is outermost and is composed of several different layers, but apparently does not absorb food as does the tegument of other parasitic helminths. The apparent lack of cuticle in the mouthless, gutless nematode parasite of *Megaselia* and its possession of microvilli suggest that nutrients pass through the body wall. The outer covering of this worm has more in common with a tapeworm tegument than has a normal nematode body wall.

The adult female parasite of *Deladenus siricidicola*[11,12] from the haemocoel of the wood wasp *Sirex noctilio*

107

has scattered clusters of microvilli, but I am not yet certain that its cuticle is entirely absent.

I thank Dr Donald L. Lee, Houghton Poultry Research Institute, for help, Dr N. W. Hussey, Glasshouse Crops Research Institute, for supplying infected *Megaselia halterata* and Mr F. Wilson of the CSIRO Sirex Unit for supplying *Deladenus siricidicola*.

[1] Nickle, W. R., *Proc. Helminth. Soc. Wash.*, **34**, 72 (1967).
[2] Welch, H. E., in *Insect Pathology* (edit. by Steinhaus, E. A.), 363 (Academic Press, New York, 1963).
[3] Welch, H. E., *A. Rev. Ent.*, **10** (1965).
[4] Poinar, G. O., and Doncaster, C. C., *Nematologica*, **11**, 73 (1965).
[5] Hussey, N. W., *Mushr. Sci.*, **4**, 260 (1959).
[6] Hussey, N. W., *Proc. Thirteenth Intern. Congr. Ent.*, London, 1964, 752 (1965).
[7] Wright, K. A., and Jones, N. O., *Nematologica*, **11**, 125 (1965).
[8] Lee, D. L., *Parasitology*, **56**, 127 (1966).
[9] Watson, B. D., *Quart. J. Microsc. Sci.*, **106**, 83 (1965).
[10] Wisse, E., and Daems, W. Th., *J. Ultrastruct. Res.*, **24**, 210 (1968).
[11] Bedding, R. A., *Nature*, **214**, 174 (1967).
[12] Bedding, R. A., *Nematologica*, **14**, 515 (1968).

Changes in the Ultrastructure
of *Nematospiroides dubius* (Nematoda)
Intestinal Cells during Development from Fourth Stage to Adult

T. P. BONNER, F. J. ETGES and M. G. MENEFEE

Introduction

The fine structure of intestinal cells of many nematode parasites, including adults of *Ascaris lumbricoides* (Kessel *et al.*, 1961; Sheffield, 1964), *Ancylostoma caninum* (Browne *et al.*, 1965; Lee, 1969), *Nippostrongylus brasiliensis* (Jamuar, 1966), and larval stages of *N. brasiliensis* (Lee, 1968) and *Trichinella spiralis* (Bruce, 1966) has been reported. Apart from the investigations of Ogilvie and Hockley (1968) and D. Lee (1969) on ultrastructural changes in the intestine of adult *N. brasiliensis* resulting from responses to host immune defenses, no studies have documented differentiation in intestinal cells at the electron microscopic level. That changes do occur in the intestine during development has been reported for *N. brasiliensis* by Weinstein and Jones (1956, 1959) and Weinstein (1966). Many of the morphological changes observed by these investigators appear to be related to transformation of *N. brasiliensis* from a non-feeding, free-living form to a feeding parasite in rats.

The present communication concerns the fine structure of *Nematospiroides dubius* intestinal cells, and changes which occur in ultrastructure during development of that parasite from fourth stage to adult.

Materials and Methods

Nematospiroides dubius was maintained by the methods of Sommerville and Weinstein (1964). Worms were recovered from the small intestine of mice at intervals. Fourth-stage worms were placed into cold 3% glutaraldehyde in 0.1 M phosphate buffer containing 0.25 M

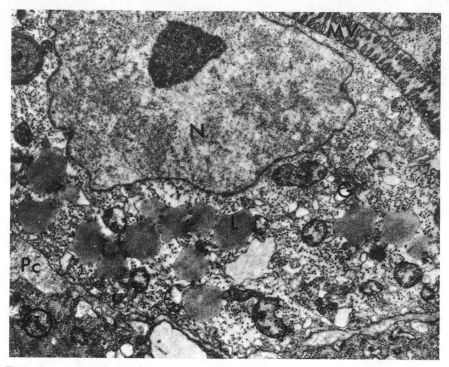

Fig. 1. Portion of intestinal cell in fourth-stage worm showing microvilli (*MV*), nucleus (*N*), Golgi body (*G*), lipid inclusions (*L*), and pseudocoelom (*Pc*). Note general distribution of mitochondria (*). × 7700

sucrose and 2.0 mM calcium chloride and cut into pieces, while adult worms were dissected and pieces of the isolated intestine were fixed in the same solution. After fixation for two hours, the specimens were washed in buffer, postfixed in 1% osmium tetroxide for 2 hours, washed in buffer, dehydrated in ethanol and embedded in Araldite 502 (Luft, 1961) or a low-viscosity medium (Spurr, 1969). Thin sections were cut with diamond knives on Sorvall MT-2 and MT-2B ultramicrotomes. Sections displaying silver interference colors were stained with uranyl acetate and lead citrate and examined in RCA EMU-3F and AEI-801 electron microscopes.

Results

General Ultrastructure

Intestinal cells of *N. dubius* displayed an ultrastructure similar to that described in other nematodes, especially *Nippostrongylus brasiliensis* (Jamuar, 1966; Ogilvie and Hockley, 1968; Lee, 1969). Therefore, only a brief description of the general fine structural features is presented. The intestine was cylindrical and in transverse section was composed of two cells. The luminal plasmlemma was folded to form microvilli (Figs. 1, 3, 4), and at the junction of cells the plasmalemmae were modified to form junctional complexes which appeared similar to septate desmosomes, particularly in the distal region (Fig. 2). Cells were often

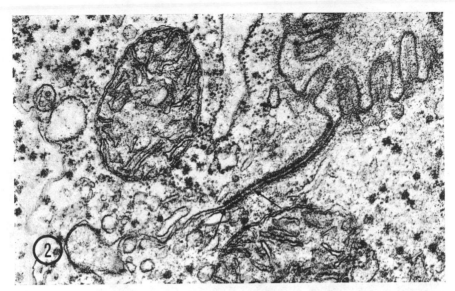

Fig. 2. Junctional complex showing separate desmosome at distal end (arrows). × 39 000

multinucleated. Large numbers of dense granules, 0.5 to 2.5 μ in diameter, were present in the cytoplasm (Fig. 3), and in living worms these pigment granules showed a reddish-brown color. Some granules with alternating light and dark bands were observed, and others were seen in which an outer dark zone surrounded a lucid, granular inner core (Fig. 4). Lipid inclusions were also present (Fig. 1).

Developmental Changes

Intestinal cells of fourth-stage worms displayed a general fine structure as described above (Fig. 1). The microvilli measured 0.4 μ in length and 0.08 μ in width. Nuclei were typically situated in the apical to mid region of cells and contained a single nucleolus and little condensed chromatin. Free ribosomes, both single and in clusters, were prominent in the cytoplasm, and a few circular to elongate rough endoplasmic reticulum (RER) cisternae were distributed in cells. Golgi complexes were also evident. Elongate to ovoid mitochondrial profiles were generally distributed throughout the cytoplasm.

Differences in this ultrastructure were seen in worms undergoing the molt from fourth to fifth (adult) stage (Fig. 3). Most striking was the presence of large amounts of pigment granules and glycogen particles and rosettes. Nuclei were located in the apical region of cells, but contained more condensed chromatin. A paucity of RER and Golgi complexes was particularly notable. Mitochondria were no longer generally distributed, but rather concentrated in the apical cytoplasm.

The ultrastructural organization of cells in mature adult worms (3 months old) was different from that of younger worms (Fig. 4). The microvilli were

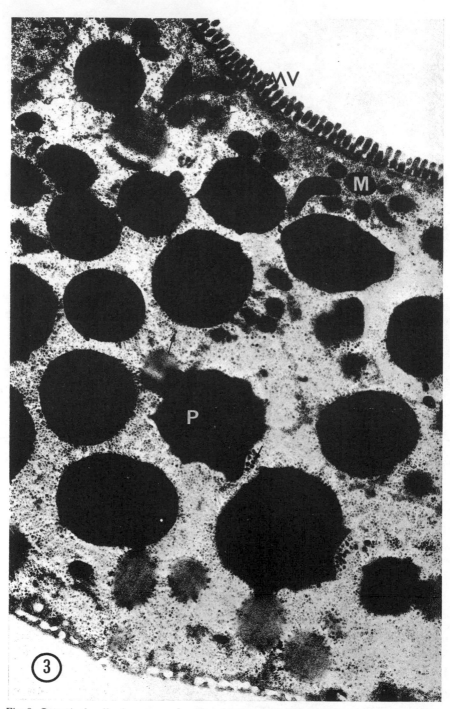

Fig. 3. Intestinal cell of worm undergoing fourth molt. Note large numbers of pigment granules (*P*), glycogen particles (arrows) apical concentration of mitochondria (*M*), and microvilli (*MV*). × 15100

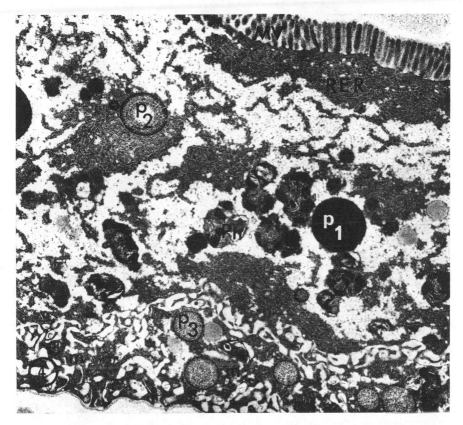

Fig. 4. Adult intestinal cell displaying rough endoplasmic reticulum (*RER*), phagolysosomes (*Ph*), and basally restricted mitochondria (*m*). Note 3 types of pigment granules P_1, P_2, P_3), and microvilli (*MV*). $\times$ 12000

0.8 μ long by 0.08 μ wide. Nuclei were generally located in the mid to basal region of cells. Abundant RER in the form of parallel arrays was scattered throughout cells, particularly in the apical region and juxtanuclear zone. Golgi complexes were not observed in adult cells. Mitochondria were redistributed and generally restricted to the basal region of cells. The basal plasma membrane was highly folded and extended well into cells.

Although pigment granules were still prominent, they were not as numerous as in younger worms (Fig. 4). In their place, many pleomorphic inclusions were distributed throughout the cytoplasm. These organelles were 0.5 to 3.0 μ in diameter, limited by a single membrane, and contained numerous membranous elements (Fig. 5) and varying amounts of electron dense, granular deposits (Fig. 6). The deposits had a particulate substructure like that of pigment granules. These structures are termed phagolysosomes.

113

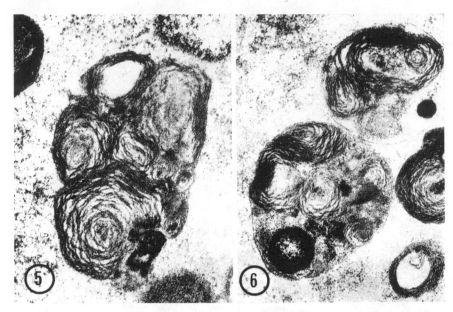

Fig. 5. Phagolysosome containing membrane elements. × 43500

Fig. 6. Phagolysosomes containing electron dense deposits. × 33000

Discussion

During development from the third-stage larva to mature adult, *N. dubius* changes from a non-feeding, free-living organism to a parasite of the mouse intestine. The present observations demonstrated ultrastructural changes in intestinal cells as worms developed from fourth stage to mature adult. These morphological changes may reflect functional activities of the cells. One of the most conspicuous changes involved pigment granules and phagolysosomes. Weinstein and Jones (1956, 1959) observed pigment granule formation in *Nippostrongylus brasiliensis*. Small, colorless, birefringent granules in the intestinal cells developed into pigment granules when larvae were grown in culture media containing hemoglobin. The formation of pigment granules under in vitro conditions was essentially the same in morphology as that seen by those investigators *in vivo*. The colorless granules accumulated neutral red, displayed a strong acid phosphatase reaction, and were identified as primary lysosomes (Weinstein, 1966). Electron microscopy revealed that the pigment granules were finely granular and enclosed by a single membrane (Weinstein, 1966), and agreed with an earlier report by Jamuar (1966). The pigment granules of *Nematospiroides dubius* were like those of *Nippostrongylus brasiliensis* and considering the parallel developmental patterns of these closely related organisms it is likely that pigment granules form in a similar manner. No reports have documented changes in pigment granules of *N. brasiliensis* after their formation. Present observations suggest that pigment granules of *Nematospiroides dubius* intestinal cells were degraded within phagolysosomes. Since the material comprising pigment granules is most likely derived from host hemoglobin, transformation of hemoglobin-containing vacuoles into phagolysosomes may be similar to that described in hepatic cells by Goldfischer et al. (1970). The possibility that nutrient is released by this process can only be suggested, but intracellular digestion of a similar pattern is well documented in other cell systems (De Duve, 1963; Novikoff, 1963).

Organelles involved in protein synthesis also underwent a marked morphogenesis. In fourth-stage worms intestinal cells contained large numbers of free ribosomes and sparsely developed RER. This feature is similar to that of "retaining cells" which generally do not secrete products of protein synthesis (Birbeck and Mercer, 1961). In contrast, intestinal cells of adult worms possessed well developed RER, a morphological indication of protein secretion at this time.

Another interesting reorganization within intestinal cells involved the distribution of mitochondria which is undoubtedly related to changing energy requirements of these cells, but the precise physiological significance of this morphogenetic change is not yet apparent.

References

Birbeck, M., Mercer, E.: Cytology of cells which synthesize protein. Nature (Lond.) **189**, 558–560 (1961).

Browne, H., Chowdhury, A., Lipscomb, L.: Further studies on the ultrastructure and histochemistry of the intestinal wall of Ancylostoma caninum. J. Parasit. **51**, 385–391 (1965).

Bruce, R.: The fine structure of the intestine and hindgut of the larvae of Trichinella spiralis. Parasitology **56**, 259–265 (1966).

De Duve, C.: The lysosome concept. In: Symposium on lysosomes, pp. 1–31, edit. by A. de Rueck, M. Cameron. Boston: Little, Browne and Co. 1963.

Goldfischer, S., Novikoff, A., Albala, A., Biempica, L.: Hemoglobin uptake by rat hepatocytes and its breakdown within lysosomes. J. Cell Biol. **44**, 513–530 (1970).

Jamuar, M.: Cytochemical and electron microscope studies on the pharynx and intestinal epithelium of Nippostrongylus brasiliensis. J. Parasit. **52**, 1116–1128 (1966).

Kessel, R., Prestage, J., Sekhon, S., Smalley, R., Beams, H.: Cytological studies on the intestinal epithelial cells of Ascaris lumbricoides suum. Trans. Amer. micr. Soc. **80**, 103–118 (1961).

Lee, C.: Ancylostoma caninum: Fine structure of the intestinal epithelium. Exp. Parasit. **24**, 336–347 (1969).

Lee, D.: The ultrastructure of the alimentary tract of the skinpenetrating larva of Nippostrongylus brasiliensis (Nematoda). J. Zool. (Lond.) **154**, 9–18 (1968).

— Changes in adult Nippostrongylus brasiliensis during development of immunity to this nematode in rats. 1. Changes in ultrastructure. Parasitology **59**, 29–39 (1969).

Novikoff, A.: Lysosomes in the physiology and pathology of cells: Contributions of staining methods. In: Symposium on lysosomes, pp. 36–77, edit. by A. de Rueck, M. Cameron. Boston: Little, Browne and Co. 1963.

Ogilvie, B., Hockley, D.: Effects of immunity on Nippostrongylus brasiliensis adult worms: Reversible and irreversible changes in infectivity, reproduction, and morphology. J. Parasit. **54**, 1073–1084 (1968).

Sheffield, H.: Electron microscope studies on the intestinal epithelium of Ascaris suum. J. Parasit. **50**, 365–379 (1964).

Sommerville, R., Weinstein, P.: Reproductive behavior of Nematospiroides dubius in vivo and in vitro. J. Parasit. **50**, 401–409 (1964).

Spurr, A.: A low-viscosity epoxy resin embedding medium for electron microscopy. J. Ultrastruct. Res. **26**, 31–43 (1969).

Weinstein, P.: The in vitro cultivation of helminths with special reference to morphogenesis. In: Biology of parasites, pp. 143–154, edit. by E. Soulsby. New York: Academic Press. 1966.

— Jones, M.: The in vitro cultivation of *Nippostrongylus muris* to the adult stage. J. Parasit. **42**, 215–236 (1956).

— — Development *in vitro* of some parasitic nematodes of vertebrates. Ann. N. Y. Acad. Sci. **77** (2), 137–162 (1959).

Ultrastructure of the Hypodermis during Cuticle Formation in the Third Molt of the Nematode *Nippostrongylus brasiliensis**

THOMAS P. BONNER and PAUL P. WEINSTEIN**

The nematode body wall consists of a cuticle which is underlain by and intimately associated with the hypodermis. The hypodermis expands into the body cavity to form dorsal, ventral and two lateral chords. The chords are connected by thin interchordal regions of tissue. Hypodermal nuclei are restricted to the chords, and mitochondria, Golgi complexes, and endoplasmic reticulum are largely concentrated in the chords (Lee, 1966; Jamuar, 1966; Bonner et al., 1970).

Several recent ultrastructural and cytochemical studies (Bird and Rogers, 1965; Kan and Davey, 1968; Samoiloff and Pasternak, 1969; Johnson et al., 1970) indicated that the cuticle is synthesized by the hypodermis. The ultrastructure of the hypodermis during formation of adult cuticle in *Nematospiroides dubius* (Bonner et al., 1970) and *Nippostrongylus brasiliensis* (Lee, 1970) is consistent with data that nematode cuticle contains large amounts of collagenous protein (Bird and Bird, 1969). In these organisms, the ultrastructure of the hypodermis resembled that of fibrogenic cells (reviewed by Ross, 1968).

The present communication describes the hypodermis during the third molt in *N. brasiliensis*. This molt was selected because it is an important event resulting in the transformation of free-living to parasitic larva. Further, corroborating biochemical data on cuticle synthesis during the third molt (Bonner et al., 1971) can be correlated with morphological events.

* This investigation was supported, in part, by awards 1-F02-AI3750-02 and AI-09625 from the National Institutes of Health, United States Public Health Service.

** We are grateful to Lida Petruniak and Dora Lou for technical assistance.

Materials and Methods

N. brasiliensis was maintained by the methods of Weinstein and Jones (1956). Free-living, third-stage (filariform) worms were isolated from charcoal culture and parasitic lung stages were recovered at two-hour intervals between 24 and 48 hours after infection (Weinstein and Jones, 1956). Worms were cut into small pieces in fixative and processed for electron microscopy (Bonner *et al.*, 1971b). A parallel series was stained en bloc with 0.5 % uranyl acetate, and some worms were impregnated with OsO_4 (Friend, 1969).

A brief description of this phase of the life history of *N. brasiliensis* is provided for orientation of the reader. Third-stage, filariform (infective) larvae normally penetrate the skin of the host (rat) and enter the blood stream. After approximately 17–20 hours they burrow out of the pulmonary circulation into the lung parenchyma. The third molt begins soon after worms reach the lungs.

Results

In filariform larvae the chordal regions projected into the body cavity between muscle quadrants, and were connected by interchordal tissue (Fig. 1). The lateral chords appeared to consist of three sectors, two sublateral and one median. The median sector was located in the apical hypodermis, appeared to be surrounded by a sublateral sector, and contained a few smooth membrane profiles, granules, and mitochondria. In those sections examined no nuclei were seen in the median sector of filariform larvae; however, this does not mean nuclei are absent. Tight junctions joined the plasma membranes of the median and sublateral sectors (Fig. 1). The sublateral sectors displayed several mitochondria, few Golgi complexes, cisternae of rough endoplasmic reticulum (RER), vacuoles, and granules (Fig. 1). Nuclei contained large amounts of heterochromatin, and were never observed to possess nucleoli. In these infective larvae, no evidence of cuticle formation was noted.

By 24 hours after infection striking changes were observed in the hypodermal chords. Nuclei were much larger, and contained a prominent nucleolus but little condensed chromatin (Fig. 2). The cytoplasm was replete with ribosomes, and RER had increased in amount. Many irregularly shaped and branched mitochondria were evident, and comparatively few Golgi complexes were seen at this time.

As development proceeded the majority of ribosomes became associated with the proliferating ER, mitochondria displayed more ovoid profiles, and Golgi complexes were more numerous. The first signs of cuticle formation were noted at this time (approximately 28 hours after infection). The proliferation of these organelles continued and many small vesicles, some of the "coated" variety, appeared in the apical cytoplasm along with multivesicular bodies (Fig. 3). The median sector was distinguished from sublateral sectors by its location and elaborately developed Golgi complexes (Fig. 3). These Golgi complexes contained 12–20 lamellae with associated vesicles, whereas, in the sublateral sectors Golgi complexes displayed 2–5 lamellae. Nuclei were observed in the median sector of lung-stage worms. At this time (36 hours after infection) the newly forming cuticle is clearly recognized.

When worms were incubated in OsO_4 according to the methods of Friend (1969) many, but not all, of the vesicles contained deposits (Fig. 4). Deposits were also seen in some of the vesicles within multivesicular bodies (Fig. 5), Golgi complexes, and lipid droplets.

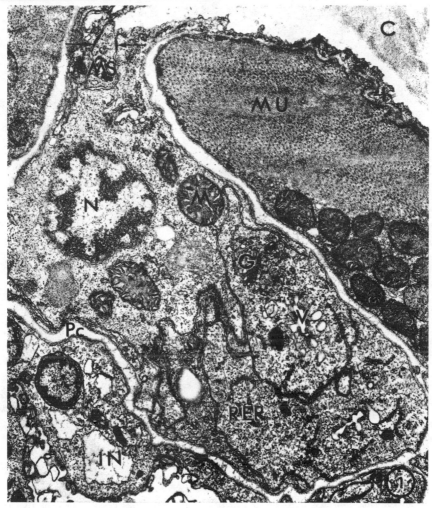

Fig. 1. Transverse section through filariform larva showing lateral chord and interchordal zone (*I*), above muscle layer (*MU*). The median sector (*MS*) of cytoplasm is surrounded by and attached to a sublateral sector. Note tight junctions (arrows). The sublateral sectors contain mitochondria (*M*), rough endoplasmic reticulum (*RER*), Golgi complex (*G*), vacuoles (*V*), and nucleus (*N*). Note cuticle (*C*), pseudocoelom (*Pc*) and intestine (*IN*). × 19000

From 36 through 54 hours after infection the fourth-stage cuticle to increase in size and finally attained the structure seen in mature fourth-stage worms. During this time the structure of the hypodermis was similar to that described above for worms at 36 hours after infection.

The interchordal cytoplasm also developed from a stage containing only a few smooth membranes to one displaying RER, mitochondria, vesicles and multi-vesicular bodies (Figs. 1, 6). The appearance of these organelles was greatest near

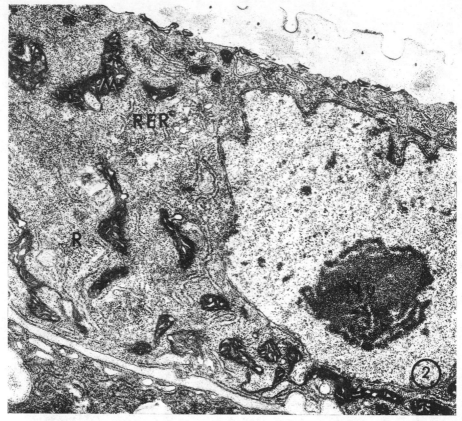

Fig. 2. Longitudinal section through hypodermis at 24 hours after infection showing nucleus with nucleolus (*Nu*), mitochondria (*M*), ribosomes (*R*) and rough endoplasmic reticulum (*RER*). × 19000

the chords and decreased in the center of the interchordal region. Details of the structural formation of cuticle during the third molt will be reported in another communication.

Discussion

The third molt in *N. brasiliensis* is recognized as the first parasitic molt, and as such is regarded as a developmental step in transition of the free-living to the parasitic organism. Associated with this transition were changes in the ultra-structural organization of the hypodermis which reflect synthesis of the cuticle of the first parasitic stage. These changes involved the organelles which function in protein synthesis.

Recent studies have demonstrated that third-stage *N. brasiliensis* was able to absorb ^{14}C-proline and utilize that amino acid for synthesis of cuticular protein during the third molt. Additionally, ^{14}C-proline was converted to ^{14}C-hydroxy-

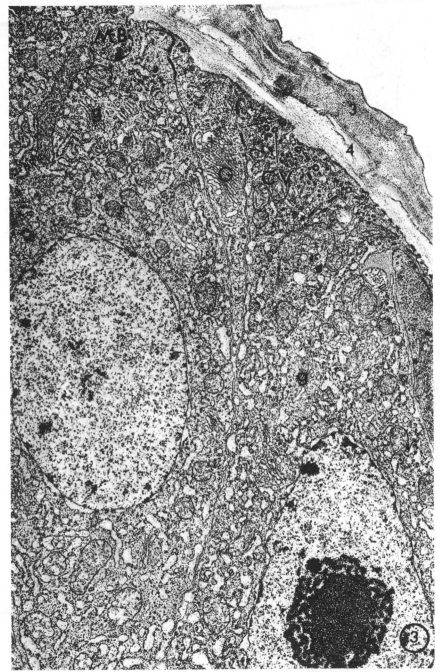

Fig. 3. Oblique section through lateral hypodermal chord at 36 hours after infection. The median sector contains a large Golgi complex (G) while smaller Golgi complexes (g) are seen in the sublateral sectors. Note elaborately development RER, mitochondria, multivesicular body (MB), coated vesicles (CV), third-stage cuticle (3) and forming fourth stage cuticle (4). × 16500

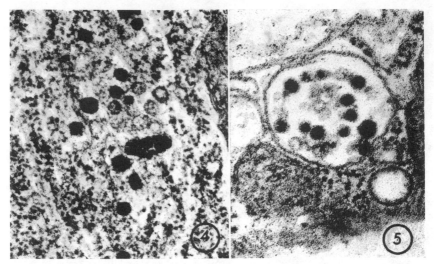

Fig. 4. Portion of hypodermis illustrating deposition of OsO$_4$ in small vesicles. $\times$ 70000

Fig. 5. Portion of hypodermis illustrating deposition of OsO$_4$ in vesicles within multivesicular body. $\times$ 110000

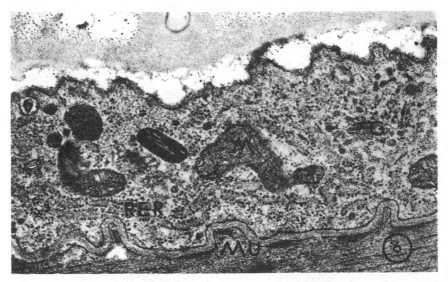

Fig. 6. Longitudinal section of interchordal cytoplasm near lateral chord at 30 hours after infection. Note *RER*, mitochondrion (*M*), Golgi complex (*G*). Muscle cell (*Mu*). Compare with interchordal region in Fig. 1. $\times$40400

proline (Bonner *et al.*, 1971). These observations, along with an earlier report on the amino acid composition of *N. brasiliensis* fourth-stage cuticle (Simmonds, 1958), indicate that the cuticle contains collagenous protein.

The fine structure of the hypodermis seen in the present study was particularly consistent with other fibrogenic cell systems with respect to the development and form of the RER, Golgi complexes, and vacuole systems. Also, the ultrastructure of the hypodermis during the third molt of *N. brasiliensis* shares many common features of the hypodermis of *N. brasiliensis* (Lee, 1970) and *N. dubius* (Bonner et al., 1970) during the fourth molt.

One noticeable difference between hypodermal ultrastructure reported in the present study and that of the fourth molt of *N. brasiliensis* and *N. dubius* is the degree of development of the Golgi complexes. Golgi complexes did not appear to be as prominent in the hypodermis of fourth-stage *N. dubius* (Bonner et al., 1970) or *N. brasiliensis* (Lee, 1970) as they were in hypodermis of third-stage *N. brasiliensis*. The appearance and elaborate development of Golgi complexes seen in the present study must certainly reflect a role of the Golgi system in cuticle formation. When worms were incubated in OsO_4, the Golgi complexes multivesicular bodies and small vesicles displayed electron dense deposits. A similar observation by Friend (1969) on the rat epididymis suggested that these elements share a common origin. It is not yet known whether the small vesicles which contain electron dense deposits, and which may have originated from the Golgi complex, contain cuticular precursors as no evidence was obtained to suggest the release of the dense deposits from the vesicles to the forming cuticle. Other vesicles in the hypodermis did not accumulate OsO_4 suggesting (1) that two types of vesicles with different origins are present in the hypodermis, and (2) that all vesicles have a common origin but their contents may undergo alterations which render them nonreactive with OsO_4. Whatever the case, the function of the Golgi complex in cuticle formation is not clear.

Numerous invertebrate groups including nematodes, annelids (Coggeshall, 1966; Rudall, 1968), and pogonophorans (Gupta and Little, 1970) possess a collagenous cuticle. The cells which produce this cuticular collagen are of particular interest because they are epidermal in nature. In addition to recognizing the possibility that cuticle collagen is synthesized by epidermal cells in annelids and nematodes, Fitton-Jackson (1968) suggested that cuticle collagen may be produced by fibroblasts which underlie the epidermis (hypodermis). No such cells are known in nematodes, and it is clear that collagen formed during the third molt of *N. brasiliensis* is of hypodermal (epidermal) origin.

Another feature of the hypodermis which requires brief discussion is the presence of distinct cytoplasmic sectors. Lee (1966) described the hypodermis, exclusive of nerve fibers and excretory ducts, as a syncytium. In the present study, three cellular regions have been described in lateral chords. Although these sectors may be multinucleated, the hypodermis is not a single continuous cytoplasmic zone. No cellular boundaries were observed in the interchordal, dorsal or ventral chordal regions of the hypodermis.

References

Bird, A., Bird, J.: Skeletal structures and integument of Acanthocephala and Nematoda. In: Chemical Zoology, vol. III, p. 253–288, ed. by M. Florkin, B. Scheer. New York: Academic Press 1969.
— Rogers, G.: The ultrastructure of the cuticle and its formation in *Meloidogyne javanica*. Nematologica **2**, 224–230 (1965).

Bonner, T., Etges, F., Menefee, M.: Changes in the ultrastructure of *Nematospiroides dubius* (Nematoda) intestinal cells during development from fourth stage to adult. Z. Zellforsch. **119**, 526–533 (1971b).

— Menefee, M., Etges, F.: Ultrastructure of cuticle formation in a parasitic nematode, *Nematospiroides dubius*. Z. Zellforsch. **104**, 193–204 (1970).

— — Saz, H.: Synthesis of cuticular protein during the third molt in the nematode *Nippostrongylus brasiliensis*. Comp. Biochem. Physiol. **40**, 121–129 (1971a).

Coggeshall, R.: A fine structural analysis of the epidermis of the earthworm, *Lumbricus terrestris* L. J. Cell Biol. **28**, 95–108 (1966).

Fitton-Jackson, S.: The morphogenesis of collagen. In: Treatise on collagen, vol. 2(B), p. 1–66, ed. by B. Gould. New York: Academic Press 1968.

Friend, D.: Cytochemical staining of multivesicular body and Golgi vesicles. J. Cell Biol. **41**, 269–279 (1969).

Gupta, B., Little, C.: Studies on Pogonophora. 4. Fine structure of the cuticle and epidermis. Tissue and Cell **2**, 637–696 (1970).

Jamuar, M.: Electron microscope studies on the body wall of the nematode *Nippostrongylus brasiliensis*. J. Parasit. **52**, 209–232 (1966).

Johnson, P., Gundy, S. van, Thomson, W.: Cuticle formation in *Hemicyclophora arenaria*, *Aphelenchus avenae* and *Hirschmanniella gracilis*. J. Nematol. **2**, 59–79 (1970).

Kan, S., Davey, K.: Molting in a parasitic nematode, *Phocanema decipens*. III. The histochemistry of cuticle deposition and protein synthesis. Canad. J. Zool. **46**, 723–727 (1968).

Lee, D.: An electron microscope study of the body wall of third-stage larva of *Nippostrongylus brasiliensis*. Parasitology **56**, 127–135 (1966).

— Moulting in nematodes: The formation of the adult cuticle during the final moult of *Nippostrongylus brasiliensis*. Tissue and cell **2**, 139–153 (1970).

Ross, R.: The connective tissue fiber forming cell. In: Treatise on collagen, vol. 2(A), p. 2–82, ed. by B. Gould. New York: Academic Press 1968.

Rudall, K.: Comparative biology and biochemistry of collagen. In: Treatise on collagen, vol. 2(A), p. 83–137, ed. by B. Gould. New York: Academic Press 1968.

Samoiloff, M., Pasternak, J.: Nematode morphogenesis: fine structure of the molting cycles in *Panagrellus silusiae* (de Man 1913) Goodey 1945. Canad. J. Zool. **47**, 639–643 (1969).

Simmonds, R.: Studies on the sheath of fourth-stage larvae of the nematode *Nippostrongylus brasiliensis*. Exp. Parasit. **7**, 14–22 (1958).

Spurr, A.: A low-viscosity epoxy resin embedding medium for electron microscopy. J. Ultrastruct. Res. **26**, 31–43 (1969).

Weinstein, P., Jones, M.: The *in vitro* cultivation of *Nippostrongylus muris* to the adult stage. J. Parasit. **42**, 215–236 (1956).

124

Ultrastructural Investigation of the Melanization Process in *Culex pipiens* (Culicidae) in Response to a Nematode

GEORGE O. POINAR, JR. AND RUTH LEUTENEGGER

Encapsulation and melanization are two immune conditions that may occur in insects in response to an invasion by metazoan parasites. The former process involves the accumulation of host cells (generally blood cells) around the parasite, enclosing it in a so-called capsule. The latter process involves the deposition of a dark pigment, generally assumed to be melanin, on the parasite after it has reached the hemocoel of the host. These processes often occur together and have been discussed in relation to metazoan parasites of insects (6) as well as helminth parasites of arthropods (4). In discussing melanization, Salt (6) felt that this process occurred only in association with cellular encapsulation or with a reaction involving blood cells. Indeed, melanization is very commonly associated with cellular encapsulation, as was demonstrated during an ultrastructural study on the formation of a melanotic capsule in the beetle, *Diabrotica* (5). However, there are cases of melanization without a clear-cut description of an accompanying cellular response. While studying the reaction of *Aedes aegypti* larvae to the DD-136 strain of *Neoaplectana carpocapsae*, Bronskill (1) noted minute particles of melanin deposited directly on the surface of the parasite. Also, in describing the reaction of *Brugia pahangi* in *Anopheles quadrimaculatus*, Esslinger (2) observed homogeneous brown plaques intermingled with an acellular fibrous material being deposited on the surface of the nematode without any direct association with any type of host cells.

Although both of the above studies were conducted with the light microscope and it would be difficult to draw definite conclusions, they suggest that a type of humoral melanization may occur in some hosts.

The following study was undertaken to better understand the process of melanization in mosquito hosts.

MATERIALS AND METHODS

A strain of *Neoaplectana carpocapsae* Weiser collected from diseased codling moth larvae, *Laspeyresia pomonella* (L.), in Mexico by Dr. L. E. Caltagirone of this department, served as the parasite. Cultures of the host, *Culix pipiens* L., were received from Dr. R. Dadd of this department. Twenty-five late fourth-instar mosquito larvae were placed in a 6-cm diameter petri dish containing 24 ml of water. Large numbers of third stage infective juveniles of *N. carpocapsae* were added to the dish with a small amount of lactalbumin hydrolyzate to stimulate feeding activity of the mosquito larvae. The mosquito larvae were removed after 15 minutes, washed, and placed in another dish filled with water with a layer of cheesecloth 1 inch from the bottom. The cheesecloth prevented the host larvae from re-ingesting nematodes that passed through their intestine unharmed. The nematodes would fall through the cloth and settle to the bottom of the dish out of reach of the mosquito larvae.

Infected mosquito larvae were removed from the retaining dish at regular intervals and dissected in phosphate buffer; the nematodes were fixed immediately in cold 1% osmium tetroxide in phosphate buffer for 2 hours. The nematodes were then transferred to buffer, dehydrated in a series of ethanol, and embedded in Araldite 6005. Sections cut with a Porter-Blum MT-2 microtome were double stained with uranyl acetate and lead citrate and examined with an RCA-EMU-3F electron microscope at 100 kV. Pigmented sheaths removed from host larvae after 5–10 hours were placed in chloroform, benzene, acetone, ether, 95% alcohol, xylene, hydrochloric acid (38%), sulfuric acid (95%), hydrogen peroxide (30%), and sodium hydroxide (40%) for 24 hours, respectively, to determine the solubility of the pigment.

FIG. 1. Melanin sheath surrounding a *Neoaplectana carpocapsae* juvenile 2 hours after entering the body cavity of *Culex pipiens*.

FIG. 2. Definitive melanized sheath surrounding *N. carpocapsae* 7 hours after entering the body cavity of *C. pipiens*. Magnification same as Fig. 1.

FIG. 3. Lateral view of *N. carpocapsae* 2 hours after entering *C. pipiens* showing the deposition of pigment on the cuticular striations and lateral lines of the nematode.

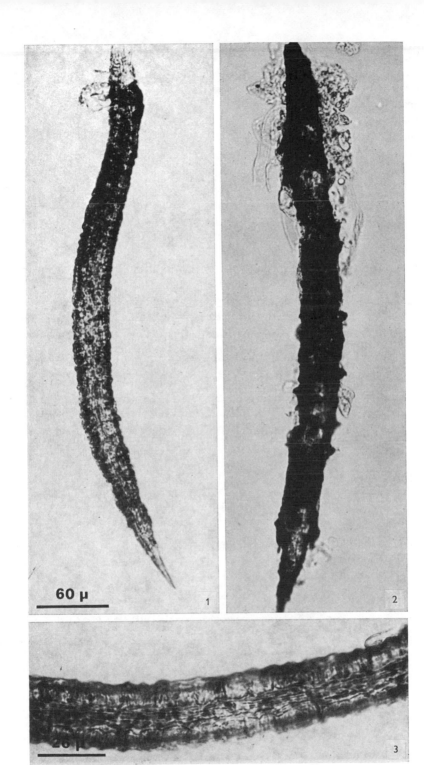

60 μ

1

2

20 μ

3

127

RESULTS

Within 15 minutes after being placed together, juvenile nematodes were observed entering the hemocoel of the host larvae from the gut. Penetration through the intestinal tract usually occurred in the prothoracic region of the mosquito and the host responded almost immediately.

Nematodes dissected from the host's hemocoel 25 minutes after the larvae were removed from the infection chamber were covered with a thin, homogeneous matrix. This deposit varied from 1 to 4 μ in thickness and made further movement difficult for the nematode. An hour after the hosts were removed from the infection chamber, a dark pigment first appeared on the surface of the parasites. This deposit was sometimes scattered evenly over the surface of the nematode, and at other times, it appeared as an anal or caudal cap or as rings on various parts of the nematode. After 2 hours, the pigment had become darker and was more or less uniformly distributed over the parasite's body (Fig. 1). During this stage, the cuticular striations and lateral lines of the nematode were finely outlined by the pigment (Fig. 3).

After 3 hours, tracheole elements were sometimes attached to the periphery of the matrix. The definitive stage was formed in 5–10 hours, when the nematode was enclosed in a strongly pigmented rigid sheath (Fig. 2). As the sheath darkened, it also became brittle and nematode movement ceased, although the parasites appeared healthy when freed from this structure.

Electron micrographs failed to reveal any definite structure or cell organelles in the initial homogeneous matrix that surrounded the nematode soon after entry into the host (Fig. 4). After being in the host for 1–2 hours, minute pigment granules began to form within this homogeneous matrix, especially in the region adjacent to the surface of the nematode (Fig. 5). These granules, which originally measured around 40–100 Å, enlarged and coalesced, and after 5–10 hours, formed a pigmented layer over the parasite (Fig. 8). Before the sheath hardened, it sometimes became detached from the body of the nematode, leaving a space which rapidly filled again with a homogeneous deposit similar to the original matrix (Fig. 6). Pigment granules appeared in this newly formed layer and a second layer of melanin was formed. This accounts for the layering effect sometimes encountered when sectioning an older sheath (Fig. 7). Each dark layer represents a ring of melanin once in contact with the cuticle of the nematode.

FIG. 4. Homogeneous matrix (H) surrounding and entering the cuticular folds (C) of *Neoaplectana carpocapsae* soon after its entry into *Culex pipiens*.

FIG. 5. Formation of minute pigment granules within the homogeneous matrix (H) adjacent to the cuticle (C) of *N. carpocapsae* 2 hours after entry into *C. pipiens*.

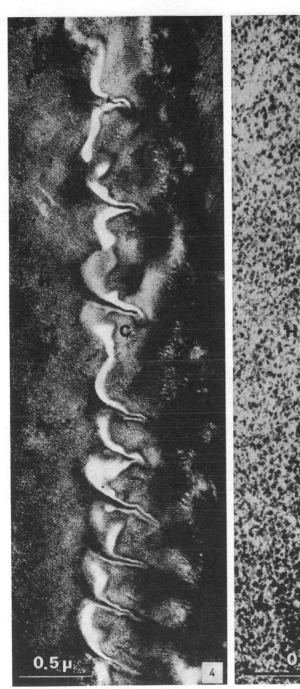

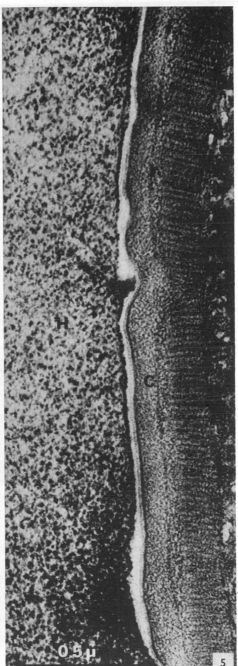

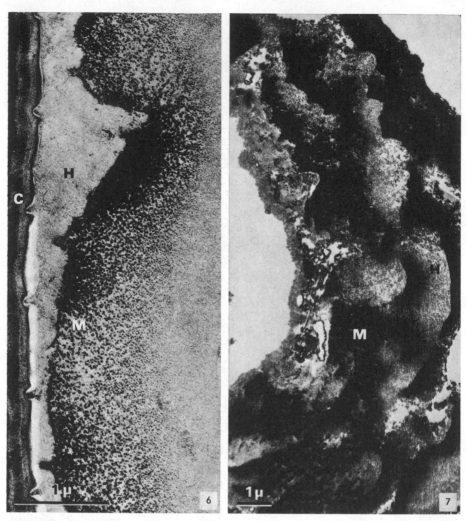

Fig. 6. Appearance of a secondary deposit of homogeneous material (*H*) adjacent to the cuticle of *Neoaplectana carpocapsae* 5 hours after entry into *Culex pipiens* when the original melanized layer (*M*) became detached.

Fig. 7. Layering effect in a 4-day-old melanized sheath showing layers of melanin (*M*) alternating with layers of homogeneous deposit (*H*).

Fig. 8. The normal definitive melanized sheath formed 5–10 hours after *Neoaplectana carpocapsae* (*N*) entered *Culex pipiens*. Note the inner melanized layer (*M*), a middle, non- or lightly melanized homogeneous region (*H*), and an outer region containing tracheole elements (*T*) and cellular debris (*C*).

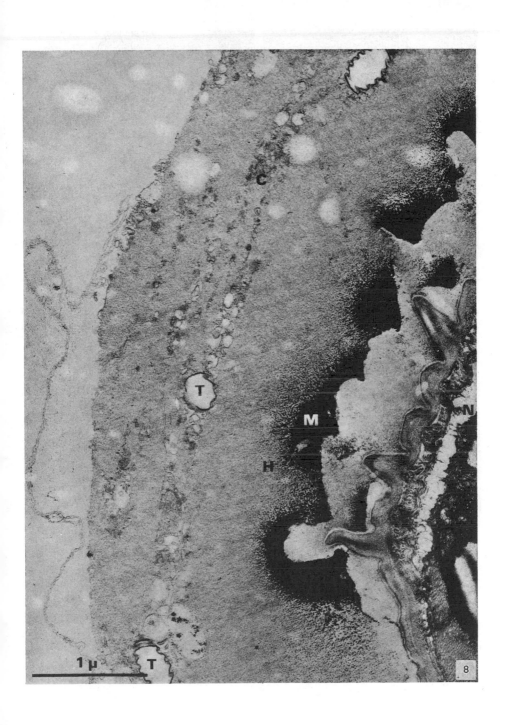

1μ

The normal definitive sheath was composed of two regions (Fig. 8). The first or inner region comprised a melanized homogeneous layer adjacent to the nematode. This was followed by a nonmelanized or lightly melanized homogeneous region which usually formed the periphery of the sheath. Occasionally, a third layer containing cellular debris and tracheole elements occurred around the homogeneous layer. Cell organelles were commonly found in this third region and probably arose from the lysis of the tracheole and blood cells. Sometimes flattened cells were also attached, however little or no melanization was observed in this area of the sheath wall.

The described defense reaction killed the nematodes in several instances, especially when just one parasite entered the host. However, when the host was attacked by two or more parasites, it frequently died since the reaction was not rapid enough to halt the escape of the bacterium, *Achromobacter nematophilus*, which is liberated from the intestine of this nematode species. The bacteria rapidly multiply in the host's hemocoel and cause a fatal septicemia, allowing the nematodes to survive and complete their development.

Tests were performed with definitive sheaths to determine the nature of the pigment. According to Fox (*3*), melanins in animal tissues are characterized by various solubility tests, some of which were conducted here. No change occurred when definitive sheaths were placed in chloroform, benzene, acetone, ether, 95% alcohol, xylene, hydrochloric acid, and sulfuric acid. The pigment was decolorized by hydrogen peroxide and partially dissolved by sodium hydroxide. All these results agree with the diagnostic tests for melanin which we will assume to be the pigmented material in this study. Melanins are a difficult group of compounds to characterize and probably exist in many forms. Fox mentions that melanin may be dissolved in concentrated sulpuric acid, and this was the case with the pigment that occurred in the capsule in *Diabrotica* (*5*). However, the material formed in the present study was not soluble in concentrated sulfuric acid, and this may reflect a difference in the chemical nature of the two pigments from the different insects.

DISCUSSION

In the present study, we have described a defense reaction in larval mosquitoes involving the formation of a pigmented sheath around invading nematode parasites. We have traced the origin of this sheath to an initial homogeneous matrix or deposit which appeared on the nematode soon after its entry into the hemocoel. It is difficult to definitely state the origin of this material without additional chemical tests. It could be connective tissue that adheres to the nematode as it brushes past the internal organs during its movement in the host. Or, it could arise from blood cells liberating this deposit on contact with the parasite. Finally, it could consist of protein coagulating out of the noncellular portion of the host's hemolymph. Our observations tend to support the latter hypothesis. It appeared to us that the deposit was too uniform to arise from chance contact of the nematode with the surface of the host's tissues. Also, if blood cells had originally made contact with the nematode, they or their inclusions would have been observed in the electron micrographs. And since the hemolymph of *C. pipiens* larvae contained relatively few blood cells, we could not account for a hemocytic origin of the deposit. Even if hemocytes liberated connective tissue some distance from the parasite, it is difficult to see how this material could migrate through the hemocoel and encircle the parasite. The only blood cells we encountered were those occasionally attached to the periphery of the deposit, and we believed they were not involved in the formation of the sheath.

Bronskill (*1*) also noted that blood cells were deposited only secondarily on the already melanized nematode in larvae of *A. aegypti*. However, we did not observe melanin particles in the cytoplasm of the hemocytes attached to the capsule as reported by Bronskill. Tracheole cells were more frequently associated with the deposit, especially after 3 hours when portions of tracheoles sometimes appeared on the periphery of the deposit. Evidence was found that these tracheal cells sometimes lysed and liberated cellular inclusions over the sheath. However, this was always after the initial deposit appeared on the nematode and was not considered essential for the melanization reaction.

Thus, we believe that the initial homogeneous deposit arose from components in the non-cellular portion of the hemolymph that precipitated or coagulated out upon contact with the parasite. This action may be triggered by a substance given off from or present on the cuticle of the nematode. This deposit then darkened through the formation and coalescence of minute pigmented particles—presumably melanin. Just how the reaction is triggered and where the chromogen and enzyme are located in this system is not known. We did not observe minute melanin particles being deposited directly on the nematode as Bronskill (*1*) described with nematodes in the hemocoel of *A. aegypti*. In the current study, a homogeneous, nonpigmented deposit always appeared on the nematode before any pigment was encountered.

On the basis of these findings, we believe that the reaction discussed here is an example of humoral melanization and that this is probably a typical response employed by members of the Culicidae against internal parasites.

This response can be distinguished from the more common hemocytic melanization where the pigment arises from the action of lysed hemocytes that make direct contact with the parasite (*5*). Humoral melanization may be common in groups of insects with few free blood cells, such as the culicids and chironomids.

Whether the chemical reactions leading to the formation of melanin are similar in the processes of hemocytic and humoral melanization is not known. It is possible that several types of melanins occur in these reactions in insects and that these may arise through different chemical pathways.

REFERENCES

1. BRONSKILL, J. F., *Can. J. Zool.* **40**, 1269 (1962).
2. ESSLINGER, J. H., *Amer. J. Trop. Med. Hyg.* **11**, 749 (1962).
3. FOX, D. L., Animal Biochromes and Structural Colours, p. 339. Cambridge Univ. Press, London and New York, 1953.
4. POINAR, G. O., JR., Arthropod immunity to worms. *In* JACKSON, G., HERMAN, R. and SINGER, I. (Eds.), Immunity to Parasitic Animals, Vol. 1, p. 173. Appleton-Century-Crofts, New York, (1969).
5. POINAR, G. O., JR., LEUTENEGGER, R. and GÖTZ, P. *J. Ultrastruct. Res.* **25**, 293 (1968).
6. SALT, G., *Parasitology* **53**, 527 (1963).

Distribution, Transmission and Experimental Procedures for Parasitic   Nematodes

Transmission of histomoniasis with male
Heterakis gallinarum (Nematoda)

By W. T. SPRINGER, JOYCE JOHNSON AND W. M. REID

That *Histomonas meleagridis* is transmitted by fully embryonated, infected *Heterakis gallinarum* ova has been established for many years. The use of surface-sterilized ova by numerous workers (Smith & Graybill, 1920; Tyzzer, 1934; Roberts, 1937; Lund, 1958; Doll, Trexler, Reynolds & Bernard, 1963; Bradley & Reid, 1966) to transmit blackhead substantiates the work of Graybill & Smith (1920), who first demonstrated this route of transmission, and indicates that the protozoan actually inhabits the heterakid ovum.

Kendall (1959) and Tyzzer (1934) presented indirect evidence of the presence of *H. meleagridis* in larval caecal worms. Swales (1948) surface-sterilized larvae and transmitted histomoniasis. Gibbs (1962) and Niimi (1937) claimed to have seen histomonads in histological sections of adult worms. Desowitz (1950) observed intracellular parasites in the gut of worms but could not positively identify them as *H. meleagridis*.

Although several workers reportedly have seen histomonads in adult worms, there are no known successful transmissions of histomonads by feeding mature heterakids containing non-embryonated eggs. The study reported herein was initiated to determine the transmissibility of histomoniasis to turkey poults by recently recovered mature *Heterakis* worms.

MATERIALS AND METHODS

Worm Source. *Heterakis gallinarum* worms, obtained from a different source for each of five trials, were collected from caeca of fowls from processing plants. The caecal contents were expressed into 500 ml beakers, diluted with Ringer's solution, and gently washed through a 100-mesh screen to remove the worms. In each study, male and female worms were placed in separate groups, washed thoroughly and, refrigerated overnight at 4 °C. Within 24 h of removal, recently killed worms were given orally to poults in all trials.

Poults. Day-old Nicholas Broad-breasted White turkey poults were used in each trial. Poults were housed in steam-sterilized Horsfall–Bauer units to maintain isolated groups. Poults were necropsied at 14 or 21 days post-inoculation, and caecal contents were examined by phase-contrast microscopy for the presence of histomonads (Kemp & Reid, 1966).

Intact worm inoculation trials. Non-viable intact adult worms were given orally with a cannula (Table 1). One group in each of the three trials was given fifty male worms per poult. Poults in a second group each received ten female worms carrying non-embryonated ova per poult (trials 1 and 2) or fifty female worms per poult (trial 3). To test for the presence of a pathogenic histomonad, ova from the same group of worms were embryonated and 1000 ova were fed to each of five poults (trial 3). In trial 2 only, caecal contents from poults inoculated with male worms were diluted with Ringer solution and inoculated rectally into another group of poults.

Triturated worm inoculation trials. Adult worms were triturated with mortar and pestle immediately before oral administration to poults (trials 4 and 5), one group in each trial receiving fifty triturated male worms. The poults in one group in trial 4 each received fifty triturated female worms, and each poult in another group received 1000 embryonated ova from the same worm source as the adult worms used in the trial.

RESULTS

Intact-worm inoculation trials. Following inoculation with fifty male worms in three trials, examination of caecal contents revealed that four out of five (4/5), 5/6 and 2/4 poults were positive for histomoniasis without lesions (Table 1). When female worms carrying non-embryonated ova were administered to poults, no histomoniasis resulted when 10, 10 and 50 worms were given in the three respective trials. All uninfected controls were negative for histomoniasis.

Upon detection of histomoniasis in trial 2, caecal contents were sub-inoculated rectally into five poults. Four of the five poults were positive 14 days post-inoculation for histomonads but no lesions appeared.

Table 1. *The transmission of* Histomonas meleagridis *with killed* Heterakis gallinarum *to day-old turkeys*

Trial no.	Treatment of groups	Duration (days)	No. positive
1	50 Intact male worms	14	4/5*
	10 Intact female worms	14	0/5
2	50 Intact male worms	20	5/6
	10 Intact female worms	20	0/9
	Uninoculated controls	20	0/5
3	50 Intact male worms	21	2/4
	50 Intact female worms	21	0/3
	Uninoculated controls	21	0/3
	Embryonated ova (1000)	14	4/4
4	50 triturated male worms	21	0/10
	50 triturated female worms	21	1/9
	Uninoculated controls	21	0/5
	Embryonated ova (1000)	14	5/5
5	50 triturated male worms	21	3/5
	Uninoculated controls	21	0/5

* No. of poults positive/no. of poults per group at end of trial.

137

Both liver and caecal lesions of histomoniasis were observed in all of four poults 14 days post-inoculation with ova derived and embryonated from the same source as worms used in trial 3.

Triturated-worm inoculation trials. Examination of caecal contents revealed 3/15 poults inoculated with fifty triturated adult male worms were positive for histomoniasis (trials 4 and 5). Histomoniasis was produced in 1/9 poults inoculated with fifty triturated female worms (trial 4). Again, no lesions were observed in poults with histomoniasis. All uninfected control poults were negative for histomonad infection. Both liver and caecal lesions of histomoniasis were observed in all of five poults 14 days post-inoculation with embryonated ova derived from the same source as worms used in trial 4.

DISCUSSION

The successful transmission of *Histomonas meleagridis* with male heterakid nematodes adds a possible new step in the life-history of this organism and lends credence to a previously postulated 'resistant' stage in the transmission of histomoniasis. Results also indicate the possibility that an intermediate non-pathogenic stage develops in the male reproductive tract and that further unknown steps may be involved in histomonad transmission by heterakid worms.

Several factors have not yet been satisfactorily explained in the transmission of *H. meleagridis*. The prolonged survival of *H. meleagridis* within the heterakid ovum, apparently in a semidormant stage, represents the only known means of survival under adverse conditions (Reid, 1967). Swales (1948) suggested after unsuccessful attempts to transmit *Histomonas* with unembryonated ova 'that the role of the caecal worm in the transmission or initiation of the disease is dependent on a living larva and that the aetiology of enterohepatitis involves factors not yet understood'. Long (1966) speculated that 'it is possible that histomonads, present within *Heterakis* larvae, are not at a stage immediately infective to birds and require further development within the larvae before becoming infective. This is suggested because *Histomonas* obtained from infected birds does not survive at low temperatures whereas the stage within *Heterakis* can survive at 4 °C for two years.' The observations of Stephenson & Hughes (1954) that lesion development occurs about 3 days later following infection with *Heterakis* ova than by rectal inoculation of free histomonads further indicates that a period of development or maturation is required for an intermediate stage.

Gibbs (1962), after observing histomonads in the vas deferens and seminal vesicles, first theorized that the reproductive tract of the male nematode becomes infected via the gut, and the female becomes infected by copulation. Further studies are needed to determine when and by what means the parasite is incorporated into the ovum following copulation.

The transmission of histomonads without disease has been previously reported by Bradley & Reid (1966) and Spindler (1967), although the mode of transmission was different. Tyzzer (1934) reported an unsuccessful attempt to transmit *Histomonas meleagridis* with male heterakid nematodes. Apparently without microscopic examination of the caecal contents, he concluded in the absence of typical blackhead lesions that transmission had not occurred.

The non-pathogenic species *Histomonas wenrichi* has been described by Lund (1963). Organisms demonstrated in the present trials were smaller (6–12 μm) and never showed more than one flagellum, a characteristic of *H. meleagridis* rather than the non-pathogenic *H. wenrichi*. It seems highly improbable that both a pathogenic and a non-pathogenic species were present. When poults were fed embryonated ova from the same source worm used in trials 3 and 4, all developed lesions typical of blackhead, indicating that a pathogenic organism was present.

Histomonads from poults in trial 2 were subpassed to susceptible poults by rectal inoculation of the infected caecal material in an attempt to induce pathogenicity. Propagation of the organism in 4/5 poults occurred with some becoming flagellated.

Throughout the study, only one poult became infected following inoculation with female heterakids. Trituration of the ova apparently liberated histomonads and permitted a low-level infection. Microscopic examination of triturated worms showed that liberation of histomonads had occurred. The reduced rate of infectivity with triturated male worms indicates that the liberated histomonads were destroyed by the low pH of the gizzard.

SUMMARY

Histomoniasis was transmitted to poults inoculated orally with recently harvested whole intact male worms in three trials (4/5, 5/6, 2/4), but not with female worms (0/5, 0/9, 0/3). When triturated male worms were given in two trials, the rate of transmission was reduced (3/5, 0/10). Triturated female worms given orally produced histomoniasis in 1/9 poults. These results suggest that further unknown steps may be involved in histomonad transmission by heterakid worms.

This study was partially supported by NSF Grant GB-5227.

REFERENCES

BRADLEY, R. E. & REID, W. M. (1966). *Histomonas meleagridis* and several bacteria as agents of infectious enterohepatitis in gnotobiotic turkeys. *Exp. Parasit.* **19**, 91–101.

DESOWITZ, R. S. (1950). Protozoan hyperparasitism of *Heterakis gallinae*. *Nature, Lond.* **165**, 1023–4.

DOLL, J. P., TREXLER, P. C., REYNOLDS, L. I., & BERNARD, G. R. (1963). The use of peracetic acid to obtain germfree invertebrate eggs for gnotobiotic studies. *Am. Midl. Nat.* **69**, 231–9.

GIBBS, B. J. (1962). The occurrence of the protozoan parasite *Histomonas meleagridis* in the adults and eggs of the cecal worm *Heterakis gallinae*. *J. Protozool.* **9**, 288–93.

GRAYBILL, H. W. & SMITH, T. (1920). Production of fatal blackhead in turkeys by feeding embryonated eggs of *Heterakis papillosa*. *J. exp. Med.* **31**, 647–55.

KEMP, R. L. & REID, W. M. (1966). Studies on the etiology of blackhead disease: the roles of *Histomonas meleagridis* and *Candida albicans* in the United States. *Poult. Sci.* **45**, 1296–302.

KENDALL, S. B. (1959). The occurrence of *Histomonas meleagridis* in *Heterakis gallinae*. *Parasitology* **49**, 169–72.

LONG, P. L. (1966). Transmission and experimental infection Histomoniasis. *Poult. Rev.* **6**, 7–11.

LUND, E. E. (1958). Growth and development of *Heterakis gallinae* in turkeys and chickens infected with *Histomonas meleagridis*. *J. Parasit.* **44**, 297–301.

LUND, E. E. (1963). *Histomonas wenrichi* n.sp. (Mastigophora: Mastigomoebidae), a non-pathogenic parasite of gallinaceous birds. *J. Protozool.* **10**, 401–4.

NIIMI, D. (1937). Studies on blackhead. II. Mode of infection. *J. Jap. Soc. Vet. Sci.* **16**, 23–6.

REID, W. M. (1967). Etiology and dissemination of the blackhead disease syndrome in turkeys and chickens. *Expl Parasit.* **21**, 249–75.

ROBERTS, F. H. S. (1937). Studies on the life history and economic importance of *Heterakis gallinae* (Gmelin, 1790; Freeborn, 1923), the caecum worm of fowls. *Aust. J. exp. Biol. med. Sci.* **15**, 429–39.

SMITH, T. & GRAYBILL, H. W. (1920). Blackhead in chickens and its experimental production by feeding embryonated eggs of *Heterakis papillosa*. *J. exp. Med.* **32**, 143–52.

SPINDLER, L. A. (1967). Experimental transmission of *Histomonas meleagridis* and *Heterakis gallinarum* by the sow-bug, *Porcellio scaber*, and its implications for further research. *Proc. helminth. Soc. Wash.* **34**, 26–9.

STEPHENSON, J. & HUGHES, D. L. (1954). Observations on the epizootiology of entero-hepatitis (blackhead) in turkeys. *Xth Wld's Poult. Congr.*, Edinburgh, pp. 282–4.

SWALES, W. E. (1948). Enterohepatitis (blackhead) in turkeys. II. Observations on transmission by the caecal worm (*Heterakis gallinae*). *Can. J. Comp. Med.* **12**, 97–100.

TYZZER, E. E. (1934). Studies on histomoniasis, or 'blackhead' infection in the chicken and the turkey. *Proc. Am. Acad. Arts Sci.* **69**, 189–264.

Ascaridia galli (Schrank, 1788) from the Chukar Partridge, Alectoris chukar (Gray), in Nevada

F. Donald Tibbitts
and Bert B. Babero

The chukar partridge, native to Asia and southern Europe, is an introduced game bird in the western United States. The presence of an unidentified roundworm occurring in the small intestine of chukar partridges from a privately owned Nevada game farm was reported to the writers by Dr. W. G. Dale, Carson City, Nevada. Subsequent necropsies of about 250 adult birds revealed that approximately two-thirds of them harbored a nematode which we have identified as A. galli. Although Yamaguti (1961; Systema Helminthum: The Nematodes of Vertebrates) cites A. compar (Schrank, 1790) from the chukar partridge, the present paper appears to be the only published report of A. galli from this host. Autopsy records in the University of Nevada Museum of Biology and consultation with Mr. Glen C. Christensen, Game Biologist, Nevada Fish and Game Commission, revealed that Nevada chukar partridges seldom harbor helminth parasites, and there is no record of A. galli occurring in the wild avian population in the state nor in hosts at the Nevada Fish and Game Commission's Mason Valley Game Farm.

The Nevada worms, while agreeing in all important details with published descriptions of A. galli, were generally shorter than those usually taken from poultry. The males ranged from 34 to 47 mm in length and females from 53 to 77 mm compared with previously published ranges of 30 to 80 mm for males and 60 to 120 mm for females—see Cram (1927; U. S. Natl. Mus. Bull. 140: 1–465) and Schwartz (1925; J. Agr. Res. 30: 763–772). The somewhat smaller size of the partridge Ascaridia may be due to crowding in the intestine.

It is suspected that cross-transmission with A. galli occurred when partridges were housed in runs frequented by Bantam chickens. Except for some intestinal obstruction—over 260 worms were taken from one bird—no obvious pathology was observed. However, the tissue penetration phase of the life cycle of A. galli and the pathologic effects upon young fowl have been reported by numerous investigators, including Ackert and Tugwell (1948, J. Parasit. 34, Suppl.: 32).

141

SURVIVAL OF JUVENILE AND ADULT *STEPHANURUS DENTATUS* IN VITRO

Francis G. Tromba and Frank W. Douvres

Populations of selected larval and adult nematodes surviving in vitro have been commonly used as sources of antigens (Thorson, 1963). Survival, for longer than a few days to a week, is not a goal in such procedures since the incubation fluids, by design, are simple and usually not nutritive. Survivals for longer periods, where antigen collection was not the primary objective, were reviewed by Hobson (1948). Extended survival of adult nematodes has subsequently been reported by Earl (1959), Taylor (1960), Weinstein and Sawyer (1961), Weinstein et al. (1963), and von Brand et al. (1963).

In connection with the development of techniques for in vitro cultivation of *Stephanurus dentatus* from infective larvae (Douvres et al., 1966), some media supported survival of advanced stages of this nematode. Accordingly, trials were conducted with juveniles and adults. This preliminary work showed that *S. dentatus* would not survive for any appreciable time without whole serum. Therefore, only those trials which included serum in various concentrations are given in this report.

The work reported herein had the following objectives: to devise techniques for improving or increasing survival that might be useful in other studies, to observe activity and physical changes during survival, and to collect metabolic products from survivors. Assay of these metabolic products for antigens will be reported elsewhere.

MATERIALS AND METHODS

Source of materials

Worms for some of the trials were obtained from swine kidneys and associated tissues shipped to the laboratory from a packing plant. Such tissues were received 24 to 36 hr after slaughter and had been kept refrigerated for this period. Juveniles were collected from the perirenal tissues and adults from ureteral cysts. In the remaining trials, the worms were obtained from the portal veins or ureteral cysts of pigs killed at various intervals after a single dose of infective larvae (Table I). In all cases, the worms were washed free of host tissues in several changes of a balanced salt solution (Fenwick, 1939) and incubated overnight, or for several hours, at 39 C in the test medium. Active undamaged specimens were then selected for survival trials.

The media ingredients and their formulations, the vessels, and the cell cultures were identical, with one exception, to those described by Douvres et al. (1966). In some trials, NCTC 109 without vitamin B_{12} was used (Microbiological Associates, Lot No. 51537).

Preparation and handling of cultures

In all trials, an initial ratio of 1 worm to 1 ml of medium was used. The total volume was 4 ml for tubes and 10 ml for prescription bottles. As worms died, this ratio was maintained as far as possible by combining survivors from individual vessels rather than by decreasing the quantity of medium. However, in some cases, particularly toward the end of an experiment, this procedure could not be followed.

At intervals of 2 to 3 days, the contents of the culture vessel were poured into a sterile petri dish for examination. Living worms were transferred to fresh medium in another vessel and used media were stored at -20 C in bottles. Between transfers, survivors were observed in the culture vessels with an inverted microscope.

RESULTS

The data from 10 trials with various media and vessels are given in Table I. These data

TABLE I. *Survival of juvenile and adult* Stephanurus dentatus *in vitro.*

Trial	Origin of worms	Age (days)	Sex and number	Medium	Vessel	Days survival range (mean)
1	Adults,[1] exp.	322	12 M 14 F	PB-1	Tubes	3–29 (17) 3–29 (18)
9	Adults,[2] shipped	–	20 M 20 F	PB-1	Tubes	2–23 (12) 2–16 (8)
20	Juveniles, shipped	–	40 M	PB-1 and NCTC 109 v/v	Tubes	2–61 (32)
23	Adults, shipped	–	18 M 18 F	PB-1 and NCTC 109 v/v	Tubes	2–17 (7) 2–17 (10)
24	Adults, shipped	–	54 M 19 F	NCTC 109 10% serum	Tubes	2–45 (22) 2–45 (28)
30	Adults, exp.	486	7 M 7 F	KW-1 with cell culture[4]	P-4 bottles[5]	6–28 (23) 4–28 (23)
31	Juveniles, shipped	–	13 M 12 F 10 F	KW-1 with cell culture KW-1 alone	All in P-4 bottles	2–70 (36) 3–11 (9) 3–33 (18)
32	Juveniles,[3] exp.	150	13 M 22 F	KW-1 with cell culture	P-4 bottles	2–198 (73) 2–183 (51)
33	Juveniles, exp.	219	32 M 30 F	KW-1 with cell culture	P-4 bottles	1–132 (100) 1–132 (79)
35	Juveniles, exp.	249	10 M 10 F 10 M 10 F	KW-1 with cell culture KW-1 alone	All in P-4 bottles	112–115 (113) 90–115 (106) 9–112 (86) 25–103 (84)

M, male; F, female.
[1] Adults recovered from ureteral cysts of experimentally infected pig.
[2] Adults recovered from material shipped to laboratory.
[3] Juveniles recovered from portal veins of experimentally infected pig. Donors for Trials 32, 33, and 35 were littermates, infected on the same day with equal numbers of infective larvae from the same pool.
[4] Swine kidney cell cultures.
[5] Four-ounce, screw-capped prescription bottles.

were selected as representative from 42 separate experiments. At the start of the study, motility was the only objective criterion available for judging survival and condition. If motion was not seen, or could not be induced by touching the worm with a dissecting needle, death was presumed to have occurred. As the work progressed, it was found that color, integrity and functioning of the internal organs, and ability to regulate osmotic pressure, as evidenced by maintenance of normal body length, were accurate guides to condition.

In freshly isolated worms, and healthy survivors, the perienteric fluid is pink, the lateral canals are light red, the excretory glands are milky white, and the intestinal wall is black. Moribund or dead worms are less vividly colored. In some motile or otherwise apparently healthy worms, rupture of the intestine, uteri, or seminal vesicle was noted. The actual break was often not seen, but it could be inferred from observing contents of these organs in the pseudocoel. Other evidence of deteriorating condition was the inability of males to retract protruded spicules or, in both sexes but particularly in females, a shrinking or elongation of the body. When these morbid changes were recognized, the affected worms were discarded, even if motile. As it appeared likely that early deaths following damage to internal organs might be a result of trauma during isolation, worms were held for 12 to 24 hr before a final selection was made to begin an experiment. This selection was done in all except the first two trials reported in Table I.

In recently isolated healthy survivors and in some individuals keep for extended periods on cell cultures, particulate intestinal contents could be seen moving back and forth apparently in response to coiling and twisting motions of the body. Particulate material was frequently seen being ejected from the anus. In worms kept in fluid media alone, the intestine became virtually cleared of particulate material in about a week. Some of the worms in cell cultures continued to eject particulate material for much longer periods. In cell cultures with active worms, the cell layer often became patchy or lacy after several days. This condition seemed to be caused, in part at least, by burrowing motions of the worms during which the head was bent toward the cell layer and twisted repeatedly. Very little change was seen in the excretory glands of healthy survivors. In an occasional individual, the glands appeared less opaque, but material was only rarely seen passing from the excretory pore. Regurgitation of material from the mouth was frequently seen, particularly in recent isolates, and had to be differentiated from excretions through the pore.

A striking change was observed in the coelomocytes of all worms surviving in media containing NCTC 109. As early as 7 days after isolation, these cells became faintly pink. The color gradually intensified until, after several weeks, it was bright red. Such changes did not occur in other media or in formulation of NCTC 109 without vitamin B_{12}.

Adult females deposited large numbers of fertile eggs for 2 to 3 days after isolation. On several occasions these were recovered and cultured to normal infective larvae. For a short time after this a few infertile eggs were found, but no eggs were seen after 4 to 5 days.

Several conclusions relative to survival can be drawn from the data in Table I. Although fewer trials were made with them, adults generally did not survive as well as juveniles. It is also apparent that juveniles recovered from the perirenal tissues (trial 31) did not survive as long as those recovered from the portal veins (trials 32, 33, 35). Delay between death of the host and recovery of the worms is another obvious variable and can be expected to decrease survival. Overall, KW-1 was superior to other fluid media and KW-1 with cell cultures was the best of all combinations employed.

Except for the worms in cell cultures of trial 35, the data in Table I do not reflect the stability of the surviving populations. For example, the population of males in KW-1 alone (trial 35) remained stable from the 20th to 57th day. Similarly, females in KW-1 alone had no losses from the 25th to 51st day. In trial 32 there was a 39-day period without a death, and in trial 33 one of 58 days.

DISCUSSION

Hobson (1948) reviewed the records then extant on survival of parasitic nematodes in vitro. He concluded that blood and tissue inhabiting species survived longer, particularly in the larval stages, than did larval or adult intestinal parasites. Confirmation was given in subsequent reports on the extended survival of filariids by Earl (1959), Taylor (1960), Weinstein and Sawyer (1961), and von Brand et al. (1963), of *Angiostrongylus cantonensis* by Weinstein et al. (1963), and of *Stephanurus dentatus* in the present report. That mixed populations of larval and adult intestinal parasites may also survive for long periods in vitro has been reported by Leland and Wallace (1966) for *Cooperia punctata* and by Douvres (1966) for *Oesophagostomum dentatum*. Since all of the aforementioned nematodes can live for long periods in their normal hosts this indicates that survival in vitro is correlated with survival in vivo. Batte et al. (1960) reported that *S. dentatus* may live for 24 months in swine. Source animals maintained at this laboratory under conditions precluding reinfection have remained patent for over 3 years. This fact, added to the prepatent period, indicates a survival in excess of 4 years for *S. dentatus* in its normal host.

Eustrongylides ignotus larvae survive for years at 20 C at a very low metabolic level but at 37 C, when they are actively metabolizing, they survive for only about 5 months (von Brand and Simpson, 1942, 1944, 1945). The effect of lower temperatures on prolonging survival was not explored in the present report since the objective was to simulate "normal" conditions.

The beneficial effects of serum on growth of nematodes in vitro are well established. It is apparently an important factor in survival also. Weinstein and Sawyer (1961) found that maximum survival of *Dirofilaria uniformis* in unsupplemented NCTC 109 was 10 days, but survival was increased to 3 weeks by

144

adding serum. Weinstein et al. (1963) reported that mean survival time of *A. cantonensis* was increased from 30 days to 70 days when 10% horse serum was added to NCTC 109. In the present study, the effect of increasing concentrations of serum is not as clear-cut because of the probable contributory effects of the cell cultures. Nevertheless, it can be seen in Table I that in cultures with fluid media alone, KW-1 was superior to the others. In addition, our preliminary work showed that serum was an absolute requirement for extended survival of *S. dentatus*.

Enhanced survival in cell cultures, coupled with destruction of the cell layer, leads readily to the hypothesis that the worms ingested the cells for nourishment. This explanation was also advanced, on somewhat better grounds, by Douvres et al. (1966) to account for enhanced development of earlier stages of *S. dentatus* in vitro. In any case, the beneficial properties of the cell layer were expressed in extended survival only. None of the juveniles showed any advancement in development of the reproductive system, and egg production was not sustained for longer than 3 to 4 days.

Among other factors which may have a bearing on survival in vitro is the cumulative age of the worms. This may explain why the older adults of Trials 1 and 30 did not survive as long as the younger juveniles of Trials 32, 33, and 35. However, comparisons between trials within the same class do not confirm this. Genetic heterogeneity of host and/or parasite can also be considered a contributing factor. The fact that the donor pigs for Trials 32, 33, and 35 were littermates infected on the same day from the same pool of infective larvae is evidence against, but not a refutation of, this argument.

The concentration of pigmented material in the coelomocytes of juvenile and adult *S. dentatus* occurred only in media containing NCTC 109 formulated with vitamin B_{12}. Weinstein (1961) reported similar accumulations in coelomocytes of third-stage larvae of *Nippostrongylus brasiliensis* and *Ancylostoma caninum* when vitamin B_{12} was added to axenic cultures. Zam et al. (1963) reported that adult swine ascaris take up vitamin B_{12} in vitro. Although coelomocytes were not analyzed, vitamin B_{12} was found in associated tissues

and in the perienteric fluid. The vitamin levels in fluid tended to decrease with a concomitant rise in levels in associated tissues. This fact suggests that coelomocytes, being bathed in this fluid, probably absorb vitamin B_{12} from it. Taken together, these findings indicate a causal relationship between uptake of the vitamin and concentration of pigment in coelomocytes.

The stability of the worm populations in some of the trials indicated that they could be used as a system for experimental work. However, since the survivors showed no development and certain physiological functions ceased, they can hardly be considered normal populations. Nutritional, anthelmintic, or immunologic studies conducted with such populations should be approached, and interpreted, with these limitations in mind.

LITERATURE CITED

BATTE, E. C., R. HARKEMA, AND J. C. OSBORNE. 1960. Observations on the life cycle and pathogenicity of the swine kidney worm (*Stephanurus dentatus*). J. Am. Vet. Med. As. **136**: 622–625.

VON BRAND, T., I. B. R. BOWMAN, P. P. WEINSTEIN, AND T. K. SAWYER. 1963. Observations on the metabolism of *Dirofilaria uniformis*. Exp. Parasit. **13**: 128–133.

———, AND W. F. SIMPSON. 1942. Physiological observations upon larval *Eustrongylides*. III. Culture attempts in vitro under sterile conditions. Proc. Soc. Exp. Biol. Med. **49**: 245–248.

———, AND ———. 1944. Physiological observations upon larval *Eustrongylides*. VII. Studies upon survival and metabolism in sterile surroundings. J. Parasit. **30**: 121–129.

———, AND ———. 1945. Physiological observations upon larval *Eustrongylides*. IX. Influence of oxygen lack upon survival and glycogen consumption. Proc. Soc. Exp. Biol. Med. **60**: 368–371.

DOUVRES, F. W. 1966. In vitro growth of *Oesophagostomum dentatum* (Nematoda: Strongyloidea) from third-stage larvae to adults with observations on inhibited larval development. J. Parasit. **52**: 1033–1034.

———, F. G. TROMBA, AND D. J. DORAN. 1966. The influence of NCTC 109, serum and swine kidney cell cultures on the morphogenesis of *Stephanurus dentatus* to fourth stage, in vitro. J. Parasit. **52**: 875–889.

EARL, P. R. 1959. Filariae from the dog *in vitro*. Ann. N. Y. Acad. Sci. **77**: 163–175.

FENWICK, D. W. 1939. Studies on the saline requirements of the larvae of *Ascaris suum*. J. Helm. **17**: 211–228.

HOBSON, A. D. 1948. The physiology and cultivation in artificial media of nematodes parasitic in the alimentary tract of animals. Parasitology **38**: 183–227.

LELAND, S. E., AND L. J. WALLACE. 1966. Development to viable egg production in the rabbit duodenum of parasitic stages of *Cooperia punctata* grown in vitro. J. Parasit. **52**: 280–284.

TAYLOR, ANGELA E. R. 1960. Maintenance of filarial worms in vitro. Exp. Parasit. **9**: 113–120.

THORSON, R. E. 1963. The use of "metabolic" and somatic antigens in the diagnosis of helminthic infections. Am. J. Hyg. Monogr. Ser. (22): 60–67.

WEINSTEIN, P. P. 1961. The specific concentration of a reddish pigment in the coelomocytes of some nematodes exposed to vitamin B_{12} in vitro. J. Parasit. **47** (Suppl.): 23.

———, L. ROSEN, G. L. LAQUER, AND T. K. SAWYER. 1963. *Angiostrongylus cantonensis* infection in rats and rhesus monkeys, and observations on the survival of the parasite in vitro. Am. J. Trop. Med. Hyg. **12**: 358–377.

———, AND T. K. SAWYER. 1961. Survival of adults of *Dirofilaria uniformis* in vitro and their production of microfilariae. J. Parasit. **47** (Suppl.): 23–24.

ZAM, S. G., W. E. MARTIN, AND L. J. THOMAS, JR. 1963. In vitro uptake of Co^{60}–vitamin B_{12} by *Ascaris suum*. J. Parasit. **49**: 190–196.

DEVELOPMENT AND SURVIVAL ON PASTURE OF GASTRO-INTESTINAL NEMATODE PARASITES OF CATTLE

Aaron Goldberg

Summaries of some of our knowledge of factors affecting the development and survival of the free-living stages of gastrointestinal nematode parasites of ruminants on pasture are given by Lucker (1941), Kates (1950), Gordon (1957), Crofton (1963), and Levine (1963).

In six trials at the Agricultural Research Center, Beltsville, Maryland (Goldberg and Rubin, 1956; Goldberg and Lucker, 1959, 1963), infectiousness of different pasture plots was determined in four to six tests per trial by grazing helminth-free calves on the plots at various intervals after contaminating them with feces containing worm eggs. The trials were begun in spring, summer, and autumn. The study of development and survival of larvae within feces, under natural conditions, has been very limited. Information on development and survival of larvae within the feces, as well as infectiousness of the herbage, was obtained in the autumn trials (Goldberg and Lucker, 1963) by examination of the feces and herbage for infective larvae at intervals after deposition of the feces. In the present study, development and survival of larvae in feces and infectiousness of herbage were studied by direct examination. The trials were begun under a wider range of weather conditions than in the previous trials, and tests per trial were more numerous.

MATERIALS AND METHODS

Fourteen trials were conducted during 1960–64. Development and survival of *Ostertagia ostertagi* and *Cooperia* (*C. oncophora* and *C. punctata*) larvae were studied in 11 and nine of them, respectively, begun in spring, autumn, and winter. *Oesophagostomum radiatum* was studied in seven of them begun in spring and winter, *Haemonchus contortus* sensu lato in the three begun in July, and *Nematodirus helvetianus* in the one begun in May. A few *Trichuris* eggs were present in some trials.

In 11 of the trials, the feces containing nematode eggs were obtained from naturally infected cattle; in three, feces were obtained from experimentally inoculated calves. Batches of feces were thoroughly mixed in an electric mixer and from 7 to 18 (avg 9) pats per batch were prepared. The pats were nearly hemispherical, approximately 11 cm in diameter and 5 cm high, and had the consistency of normal cattle feces. They weighed 300 g in 10 trials, 200 g in two, and 150 and 50 g in one each. In each trial, two pats were mixed with amounts of charcoal and water considered close to optimum for larval development and were cultured in the laboratory under approximately optimal ambient temperatures, except in summer when the ambient temperature was usually above optimum. The numbers of larvae recovered from these cultures provided an index of the potential for production of infective larvae by the eggs in the pats. These cultures are referred to hereinafter

147

TABLE I. *Number of infective larvae recovered from the standard laboratory cultures made at the time pats of the same composition and weight were placed on pasture.*

Date of deposition	Wt of pats (g)	Ostertagia ostertagi	Cooperia oncophora and punctata	Oesophagostomum radiatum	Nematodirus helvetianus	Haemonchus contortus sensu lato
21– 7–60	150	—	—	—	—	1,105
22– 7–60	200	—	—	—	—	6,630
22– 7–60	50	—	—	—	—	3,445
17–10–60	300	17,325	—	—	—	—
18–10–60	300	80,210	43,190	—	—	—
16–11–60	300	312,000	—	—	—	—
16–11–60	300	15,669	485	—	—	—
2– 6–61	300	18,940	2,020	4,293	—	—
19–12–62	200	719	594	94	—	—
19–12–62	300	12,350	14,300	650	—	—
29– 4–63	300	20,658	47,032	3,595	—	—
2– 5–63	300	1,477	4,444	850	56	—
13– 3–64	300	885	585	560	—	—
17– 4–64	300	25,000	43,400	6,600	—	—

as "standard cultures." The remaining pats were dropped on permanent pastures so they would be subjected to as nearly natural conditions as possible. The pats were spaced 3 ft apart on clean, level areas that had not been grazed for at least 2 years, in the months and years indicated in Table I and Figure 1. The pastures consisted of a mixture of grasses, about 90%, and other herbs. If the height of the herbage at the beginning of a trial was greater than 6 inches, the pasture was mowed to 4 inches before the pats were deposited.

At frequent intervals, often weekly, a pat and the surrounding herbage were examined for larvae. In the trials begun in July, the herbage within a radius of 3 inches of the pat was collected; in the other trials the radius was increased to 5 inches. The herbage was clipped within 1 inch of the ground and its length measured. An average of seven such collections was made per trial. In the trials begun in December, an average of five collections was sufficient. In other trials more rapid changes in development or survival of larvae required 11 to 15 collections. The study period per trial averaged 19 weeks and ranged from 8 to 35.

Each pat collected from the pasture was weighed and examined for worm eggs. If eggs were present, a portion of the pat was cultured to determine the percentage still viable. The remainder of the pat and the herbage were placed in the Baermann apparatus on the day of collection. If many larvae were recovered, the totals were determined from dilution counts. The number of larvae present in the entire pat was computed.

The data on development and survival of larvae in the pats and on infectiousness of the herbage were examined in relation to precipitation and average temperature, which were used as indices of two major environmental factors, moisture and heat. Monthly precipitation and average monthly temperatures based on daily weather data obtained at this Research Center during the study period and for 1951–60 were obtained from U. S. Weather Bureau records (Fig. 1).

RESULTS AND DISCUSSION

Primarily because of evaporation of water, the fecal pats usually lost 25 to 55% of their weight within 1 or 2 weeks, despite the formation of a crust on the surface. After 7 weeks exposure, the weight loss of the pats deposited between March and July was more than 75%, whereas that of those deposited between October and December was 12 to 51%.

A great variety of organisms lived in close association with the pats. Annelids, dipterous and coleopterous larvae, slugs, and snails fed on the feces, and thus probably destroyed many eggs and larvae of parasitic nematodes. Isopods and adult beetles were common under the pats and in cavities formed in them by coprophagous organisms. Free-living nematodes, bacteria, and Protozoa were also important members of the community. Ants, Collembola, mites, tardigrades, and fungi were also found on the pats.

In the trials begun in July, the eggs disappeared from the pats within 1.5 weeks. Apparently, most of the eggs and emerging larvae were destroyed since preinfective larvae were rarely recovered from the feces. In the spring trials, the eggs disappeared by the 2nd to the 6th week, except for those of *Nematodirus*, which disappeared by 8 weeks, and many developing larvae were recovered in this period. In the trials begun in October, 66%, 36%, and 4% of the eggs were still present 1, 2, and 4 weeks after deposition, respectively, and many larvae were developing. In the trials begun in December, 28%, 12%, and 0.3% of the eggs were still present 4, 7, and 13 weeks after deposition, respectively. The percentage

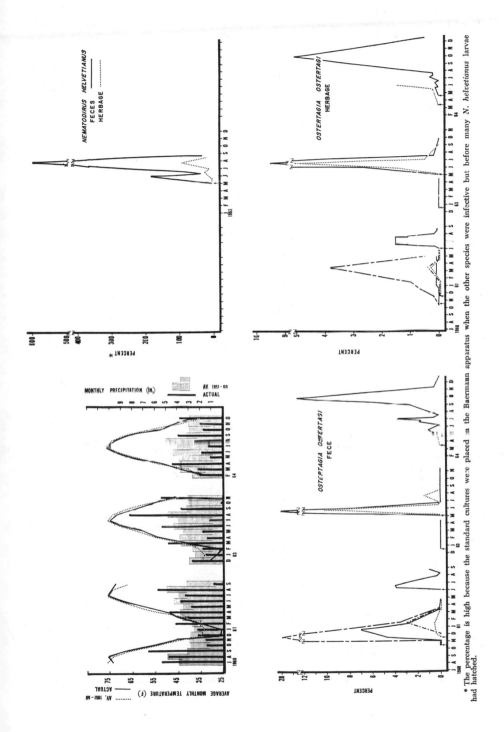

* The percentage is high because the standard cultures were placed in the Baermann apparatus when the other species were infective but before many *N. helvetianus* larvae had hatched.

149

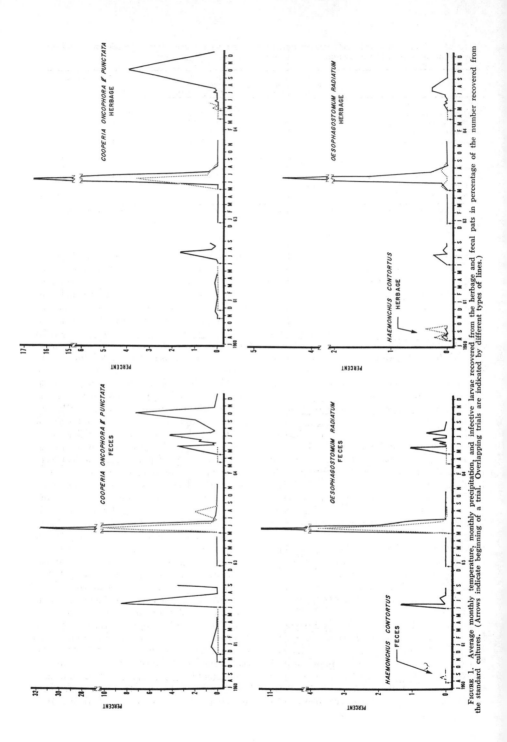

FIGURE 1. Average monthly temperature, monthly precipitation, and infective larvae recovered from the herbage and fecal pats in percentage of the number recovered from the standard cultures. (Arrows indicate beginning of a trial. Overlapping trials are indicated by different types of lines.)

150

TABLE II. *Temperature (F) and precipitation (inches) ranges for good larval development and infectiousness of herbage, and the months in which they occurred.*

	Range of monthly mean temperatures	Range of monthly precipitation	Months
	DEVELOPMENT		
Ostertagia ostertagi	55.0–70.3	1.04–3.75	May, June, Oct.
Cooperia oncophora and *punctata*	60.3–70.3	1.04–3.75	May, June
Oesophagostomum radiatum*	60.3–70.3	1.04–3.75	May, June
Nematodirus helvetianus*	60.3–69.8	1.77–5.42	May, June
	INFECTIOUSNESS OF HERBAGE		
Ostertagia ostertagi	44.0–70.3	1.63–5.42	Mar., May, June, Sept., Oct.
Cooperia oncophora and *punctata*	50.9–75.6	1.63–5.42	May, June, July, Sept., Oct.
Oesophagostomum radiatum*	69.8	5.42	June
Nematodirus helvetianus*	69.8–74.6	1.05–8.26	June, July, Aug.

* *O. radiatum* was not tested in autumn; *N. helvetianus* was present in considerable numbers only in the 2 May deposition.

of larvae recovered from pats deposited in December and cultured after 4, 7, 11, and 12 weeks of exposure was 16%, 28%, 8%, and 0.3%, respectively, of the number recovered from the standard cultures; no larvae were recovered from the pat cultured after 13 weeks on pasture. Although there was some development in the eggs during the winter, only a few of the embryos reached the vermiform stage. The only nematodes which survived the winter within the egg were *Nematodirus* and *Trichuris*.

The numbers of larvae recovered from the standard cultures are given by species in Table I. The numbers of infective larvae recovered from the herbage and fecal pats in percentage of the number recovered from the standard cultures are given in Figure 1. Table II gives, by species, the range of average monthly temperature and precipitation favorable for larval development, and for infectiousness of the herbage, and the months in which they occurred.

Where comparable, the findings generally confirm our previous studies with cattle nematodes, when calves were used to test development and survival of larvae on pasture (Goldberg and Rubin, 1956; Goldberg and Lucker, 1959, 1963).

In general, abundant moisture and moderate heat favored migration of larvae onto the herbage. For example, following deposition of pats between 29 April and 2 June, the peak recovery from the pats usually occurred in 1.5 to 3 weeks, and the peak on the herbage usually at 6 weeks. When conditions were unfavorable for migration onto the herbage, but not severe enough to destroy the larvae, many remained in the feces. For example, probably because of a deficit of 10 inches of precipitation from May to August 1964, the peak on the herbage following 17 April deposition of pats did not occur until about 25 September. Also, the peak recoveries of larvae from the pats deposited in mid-October were obtained in 4 and 7 weeks; however, because of unfavorable temperatures that prevail in late autumn and winter, the peaks on the herbage did not occur until the following March. Maximum development and survival of infective larvae occurred after deposition of pats between 17 April and 2 June, and in mid-October. Development and survival were close to zero in the trials begun in December. The limiting factor was the low temperature. The average temperature for December 1962, and January and February 1963 was 4.9 F below normal for the 3-month period. However, there was an insulating layer of snow on the ground from 21 December to 13 January and from 27 January to 6 February. When temperature and precipitation were below normal, development of infective larvae was negligible in a previous trial begun in mid-November (Goldberg and Lucker, 1963).

Ostertagia and *Cooperia* were the most abundant genera in most of the trials. Recovery of these nematodes was maximal in the trial begun on 29 April 1963. Maximum recovery from the pats, 28% *Ostertagia* and 32% *Cooperia*, was obtained 2 weeks after their deposition. The peaks on the herbage, 10%

and 17%, respectively, were attained 4 weeks later, and larvae were recovered from the herbage for 24 and 27 weeks, respectively. However, recovery from the herbage fell off rapidly during the hot dry spells of summer. Recovery of both genera was also good in the trials begun 17 April 1964 and 2 May 1963, but the peak on the herbage in the former trial was delayed to 23 weeks after deposition of the pats probably because of unusually dry conditions in the ensuing months. Good recovery of *Ostertagia* was also obtained in one trial begun in October, but peak recovery from the herbage was delayed to 21 weeks following deposition of the pats probably because of cold weather in the subsequent months. *Cooperia* was not present in that trial.

Oesophagostomum did not develop as well nor survive as long as *Ostertagia*, *Cooperia*, and *Nematodirus*. It also developed best in the trial begun on 29 April 1963, but even then the peak recovery from the herbage was only 4% of the recovery from the standard cultures, and larvae were recovered from the herbage for 10 weeks. It seems to be readily killed by extremes of heat or cold. Peak recovery from the herbage in the six other trials did not exceed 0.2%.

Only *Haemonchus* was tested in the trials begun in July. Its development and survival were poor despite approximately normal average temperatures for July, August, and September 1960, and despite the precipitation being 4.8 inches above normal for that period. The peak recoveries of larvae from the pats were only 0.02 to 0.09% of the recovery from the standard cultures, 1.5 to 5 weeks after the beginning of the trials. The peaks on the herbage were 0.06 to 0.4%, between 1.5 and 7 weeks after deposition of the pats, and larvae were recovered from the herbage for only 3.5 to 6 weeks.

In the one trial in which *Nematodirus* was present in considerable numbers, it survived 4 weeks longer than *Ostertagia*, *Cooperia*, and *Oesophagostomum*. However, it was slower in hatching and migrating onto the herbage. The peak on the herbage of the genera other than *Nematodirus* was attained in 6 to 7 weeks, whereas it was not attained by the latter until 11 weeks after deposition of the pats. The

recovery of *Nematodirus* larvae from the pats brought in from pasture was greater than that from the standard cultures because the latter were examined when the larvae of the other genera were infective but before many *Nematodirus* larvae had hatched.

The results may have been affected by the height of the herbage. After deposition of the pats, the herbage was not cut until it was collected for examination. In this region, permanent pasture plants resume appreciable growth about 1 April and provide good forage by mid-April. They become dormant about 1 November. In the present study, greatest growth occurred in spring, next greatest in summer, and least in autumn.

In two trials, the first in summer and the second in spring, Vegors (1960) compared development of gastrointestinal nematodes of cattle on pasture plots of tall and short forage. In the first trial development was the same on both lengths of forage; in the second trial more larvae were recovered from tall forage, and more worms from test calves on tall forage than on short forage. There was some difference in the effect of forage height on the different species of nematodes.

High herbage may conserve moisture near the ground and lessen direct exposure of larvae to the sun. Increased growth in the spring tends to decrease the density of larvae on the herbage; diminished growth in autumn has the opposite effect. However, in the present study, no correlation between herbage height and larvae recovered was observed, probably because the effect of herbage height was greatly overshadowed by the effect of temperature, precipitation, and the time elapsed between deposition of the pats and collection of herbage for examination.

Studies such as the present one are valuable in revealing vulnerable points in the life history of the parasites. Recommendations can then be formulated for management of grazing cattle that will result in minimum acquistion of parasites. In the present study, in general, following deposition of pats in the warm months (29 April to July) larvae were present in peak numbers in the feces between 1.5 and 5 (avg 2.5) weeks later, and in peak numbers on the herbage in 1.5 to 7 (avg 6) weeks. Following deposition in the cool months (October, November, March to 17 April), they were present in peak numbers in the feces between 4 and 23 (avg 10) weeks, and in peak numbers on the herbage in 8 to 23 (avg 17) weeks. These findings indicate that it would generally be impractical to attempt to minimize acquistion of larvae by grazing cattle, by a rotation system that is dependent on a pasture resting period to diminish the number of available larvae.

Prophylactic treatment of cattle at the beginning of the grazing season would be valuable in minimizing recontamination of pastures in the spring with species which survive the winter in the host, but not well, or at all, on pasture. Prophylactic treatment about the end of September would be most effective in minimizing contamination of pastures at the other period when conditions are optimal for development and survival on pasture. Another expedient would be to place particularly susceptible cattle, primarily young calves, on clean pastures during periods when development and survival of larvae on pasture are good.

LITERATURE CITED

CROFTON, H. D. 1963. Nematode parasite population in sheep and on pasture. Tech. Communication No. 35, Commonwealth Bureau of Helminthology, St. Albans, England.

GOLDBERG, A., AND R. RUBIN. 1956. Survival on pasture of larvae of gastrointestinal nematodes of cattle. Proc. Helm. Soc. Wash. 23: 65–68.

———, AND J. T. LUCKER. 1959. Survival on pasture of larvae of gastrointestinal nematodes of cattle. II. Spring contamination. Proc. Helm. Soc. Wash. 26: 37–42.

———, AND ———. 1963. Survival on pasture of larvae of gastrointestinal nematodes of cattle. III. Fall contamination. J. Parasit. 49: 435–442.

GORDON, H. M. 1957. Helminth diseases. Advances in Vet. Sci. 3: 287–351.

KATES, K. C. 1950. Survival on pasture of free-living stages of some common gastrointestinal nematodes of sheep. Proc. Helm. Soc. Wash. 17: 39–58.

LEVINE, N. D. 1963. Weather, climate, and the bionomics of ruminant nematode larvae. Advances in Vet. Sci. 8: 215–261.

LUCKER, J. T. 1941. Climate in relation to worm parasites of livestock, in Yearbook of Agric., USDA, p. 517–527.

VEGORS, H. H. 1960. The effect of forage height on the development of cattle nematode larvae. J. Parasit. 46 (5-Sect. 2): 39–40.

Application of the Bacteriological Pour-plate to Facilitate Mouse Pinworm Counts

Andrew Covalcine, Sei Yoshimura, and Russell F. Krueger

The following method applies the use of the bacteriological agar pour-plate to facilitate counting pinworms of mice, but it is also applicable to other intestinal nematodes. The advantages of the method are the ease of handling, uniform distribution, and the prevention of worm movement; also this method permits storage for future examination.

Mice are starved for 24 hr prior to killing. The large intestines are removed and placed in a beaker containing Bouin's fixative, which stains the worms and debris yellow and sterilizes the fecal mass. The beaker may be sealed with aluminum foil and frozen until it is convenient to prepare the plates.

Each thawed section is placed in a beaker containing 15 ml of water, cut longitudinally with scissors, and scraped of fecal material. The entire suspension is added to 20 ml of 1.5% melted agar, agitated 15 to 30 sec on a Vortex-Genie mixer, and poured into a square (100 by 100 by 20 mm) gridded disposable plastic petri dish. After the agar has solidified and cooled, the plates can be conveniently counted or stored in the refrigerator for several weeks if necessary before examination. The plates are inverted on the stage of a stereomicroscope for counting and identification. *Syphacia obvelata* and *Aspiculuris tetraptera* can be distinguished in the agar.

SEROLOGICAL REACTIONS TO INFECTION WITH
NECATOR AMERICANUS

P. A. J. BALL AND ANN BARTLETT

Man develops no complete immunity to hookworms, and people exposed to infection continue to carry worms all their lives. In the drier tropics worm loads are low, and probably reflect infrequent opportunities for infection. In wetter areas, where infection is universal and relatively heavy, worm loads are usually stable and harmless provided that dietary iron is sufficient, although a few people, and particularly young children, acquire such heavy loads as inevitably to cause disease. Patients in such areas usually become infected to their full former worm load within months of treatment with a vermifuge (GILLES, WATSON-WILLIAMS and BALL, 1964). Hookworms can live for many years, and to explain these findings one must invoke some mechanism which normally prevents the cumulation of worms in the host.

In this paper we describe some serological responses to *Necator americanus*. We have found a recurring response to first and subsequent infections, detected by a fluorescent antibody reaction, electron microscopy of larvae after incubation in serum, and complement fixation and Prausnitz-Küstner reactions; there is also a delayed response to continuing infection or repeated reinfection, detected by agglutination of coated tanned red cells.

The methods used and the results are described for each test separately.

Material

One of us (P.A.J.B.) was infected repeatedly by placing third-stage larvae (not exsheathed) on damp gauze on the back of the forearm; the eggs which appeared in the faeces were counted by the method of STOLL and HAUSHEER (1926). The numbers of larvae used and the faecal counts are shown in Figure 1. The broken line on the same graph shows the counts expected, assuming that all larvae from each challenge had matured and produced 7,000 ova per worm per day (GILLES et al., 1964), and that the average stool weight was 100 g. daily. All the serological tests were carried out with larvae resulting from this infection.

The eosinophil response and the eruption at the site of entry of the larvae differed between the first and subsequent infections. An absolute eosinophil count of 14,000 per c.mm. was attained within one week of the first infection. It fell to 2,500 during the next year, and to approximately 1,200 after 3 years, with no apparent increase following the second or subsequent infections.

The skin eruption persisted for about 3 weeks after the first and second infections, but for only one week after later infections. No larvae migrated within the skin at first infection, but after later infections most of the larvae gave rise to a creeping eruption.

Sera were also obtained from the following subjects:—

(i) 100 English adults who had never left Europe.

This work was supported by a grant from the Medical Research Council. We are grateful to Dr. L. G. Goodwin for helpful discussion, to Mr. Derek Taylor for the electron micrographs, and to many colleagues who sent us material.

(ii) More than 500 Nigerian children and adults from areas of universal infection, but with unknown egg counts.

(iii) 40 African adults from Nigeria and Kenya with known egg counts.

(iv) 16 Nigerian children aged less than 5, whose stools contained only eggs of *Ascaris* (on 3 examinations).

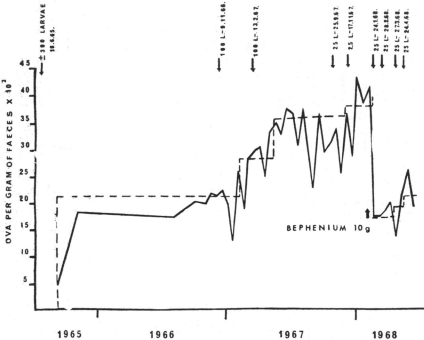

FIG. 1. Egg counts in the faeces after repeated experimental infection with *N. americanus.* Each point is the average of 4 counts made over a fortnight. The broken line shows theoretical counts, assuming maturation of all larvae, production of 7,000 eggs per worm daily, and a mean faecal weight of 100 g. daily; bephenium Alcopar (Burroughs Wellcome) 5 g. was given on 20 and 21 January, 1968

Fluorescent antibody test (FAT)

An indirect test was used, with living exsheathed larvae as the antigen. Third stage larvae were separated by the Baermann technique from charcoal cultures of faeces incubated at 28°C. for 10 days. They were exsheathed in 1 : 2,000 sodium hypochlorite in distilled water at 37°C. for 2 minutes, washed and suspended in phosphate buffered saline (pH 7·6). They were then incubated overnight at 4°C. in tubes with approximately 50 larvae in 1 ml. of 2-fold serial dilutions of serum in each. After washing they were incubated unfixed for 35 minutes at room temperature in fluorescein-labelled anti-human serum (Burroughs Wellcome), again washed 3 times and examined on slides in random order.

No precipitates could be seen by light microscopy, and larvae remained motile in all sera. In positive sera only the cuticle showed specific fluorescence; negative sera

156

showed faint non-specific fluorescence of the gut and cuticle. The reaction developed inconsistently on larvae incubated in serum for an hour; it was consistent after overnight incubation, but sometimes gave titres higher by one dilution after 48 hours. Larvae frozen immediately before incubation still gave positive reactions. Obviously damaged larvae, and larvae in sheaths, usually showed non-specific fluorescence only.

The titres attained in the experimental infection with *N. americanus* are shown in Figure 2. They began to rise a fortnight after each infection, reached a peak in about 3 months, and then slowly declined. Of 35 sera of English subjects, 20 were negative, 9 were positive 1 : 20, four 1 : 40 and two 1 : 80. Sera of infected adults from Nigeria gave titres from 1 : 20 to 1 : 1280, unrelated to worm loads.

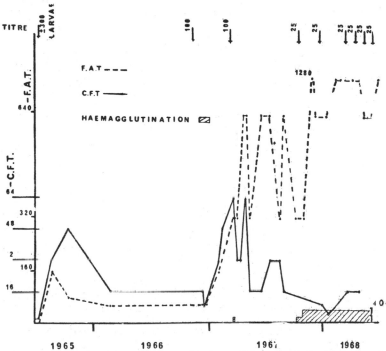

FIG. 2. Fluorescent antibody, complement fixation and haemagglutination titres following repeated experimental infection with *N. americanus*

The specificity of the antigens involved has been tested *in vitro* by using exsheathed larvae of several species of Strongyloidea, and the sera of 6 English adults, P.A.J.B. (after infection) and 2 infected Nigerian adults. The results are shown in the Table. Reactions occurred to all the species of larvae tested. In the absence of infection, titres to *N. americanus* were negative or low; *Ancylostoma duodenale* gave results intermediate between *N. americanus* and other species. A single pre-absorption, effected by diluting the serum 1 : 20 with *N. americanus* extract (1 : 10) and leaving it overnight before further dilution and addition of larvae, always removed any reaction to *N. americanus*. A positive reaction to the other species looked different from the immune reaction to *N. americanus*. The fluorescent layer was apparently specific, and was restricted to the cuticle; it was, though, thinner and less bright at all dilutions, and pre-absorption with a saline larval extract did not consistently prevent its appearance. No direct evidence could be obtained of the specificity of the acquired response to *N. americanus*, as serum obtained before the first experimental infection was no longer available.

Reciprocal titres to an indirect FAT with third stage larvae of various Strongyloidea as antigen. The sera were from 6 uninfected English adults, 1 English subject after infection with *N. americanus*, and 2 infected adult male Nigerians. The method of pre-absorption is described in the text

	N. americanus	pre-absorbed *N. americanus*	*Oesophagostomum* sp. "	*D. viviparus* "	*A. duodenale*	*A. caninum*	pre-absorbed *N. americanus* "	adult *Ascaris* cuticle "	*Oesophagostomum* sp.	pre-absorbed *N. americanus* "	*Dictyocaulus viviparus*	*Nippostrongylus brasiliensis*
European:—	0	0			40	160			80		160	160
	0	0			80	160	80	80	80	0		80
	0	0			0	160	40	80	40	40		160
	0	0			80	160	80	80	320	40		160
	0	0				40					80	
	20	0				80					80	
P.A.J.B. { Pre–	0	0										
Post–	1280	0	0	320	640	1280	160	1280	320	0	160	320
Nigerian:—	160	0				40	0		80	0		640
	320	0				40	0		160	40		640

Electron microscopy

Exsheathed larvae (of *N. americanus* only) were fixed with 5% glutaraldehyde in McIlvaine's buffer (pH 7·0) and then with Zetterqvist's osmium tetroxide. They were embedded in Araldite, and sections were examined without further staining. Larvae were examined unincubated, incubated in buffer alone, and after incubation in 6 sera (diluted 1 : 20) which were positive to FAT, 6 negative sera, and 1 negative and 4 positive sera after pre-absorption as described for the FAT.

All but one of the samples of larvae incubated in positive sera showed precipitates on the surface of the cuticle (Figures 3a and b); no precipitates were seen on larvae incubated in negative or pre-absorbed sera.

158

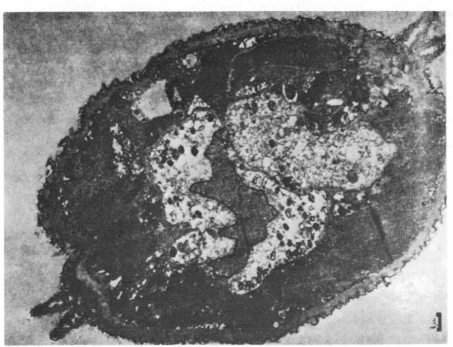

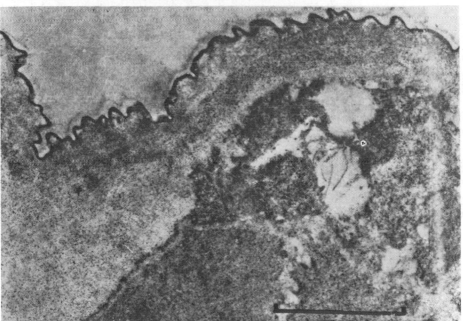

FIG. 3 (a). Electron micrograph of a cross section of an exsheathed third stage larva of *N. americanus* incubated in serum positive to the FAT showing a precipitate on the surface of the cuticle: × 3,000. (b): Cuticle of larva incubated in positive serum: × 20,000. Fixed with glutaraldehyde and osmium tetroxide.

Complement fixation test (CFT)

The antigen used was an extract of whole third-stage larvae. The larvae were stored at $-70°$C. until needed, and then ground in a glass tissue grinder in 10 volumes of veronal buffer (pH 7·6). The suspension was centrifuged for 5 minutes at 5,000 r.p.m. and the supernatant made up to a final dilution of 1 : 40. The antigenic component was shown by ultracentrifugation to be soluble. Sera, antigen and complement were incubated together at 37°C. for one hour before adding sensitized cells, and the plates were left for one hour at 37°C. and then overnight at room temperature before reading.

The titres attained in response to the experimental infection are shown in Figure 2. They reached a peak about 6 weeks after infection or reinfection and fell during the next month. With 35 English sera negative results were obtained in 34, and a titre of 1:4 in one. In Nigerian adults titres varied from negative to 1 : 64. FAT and CFT titres were not closely related, though a positive CFT was always accompanied by a positive FAT. Of the 16 children whose stools contained only ova of *Ascaris*, 14 were negative to both CFT and FAT and 2 were positive to both. 6 adult sera, with titres ranging from negative to 1 : 64, were tested with both *N. americanus* and *A. caninum* extracts as antigen; the results were identical, showing complete cross-reaction.

"Reagins"

Prausnitz-Küstner tests were carried out on serum obtained before and at intervals after the first and second experimental infections. 0·1 ml. of serum diluted 1 : 25 in normal saline was injected intradermally into the flexor aspect of the forearm of 4 subjects, with serum heated to 60°C. for one hour as a control on the other side. After 72 hours 1 : 10 *Necator* extract was pricked through the skin over the sites of injection, and the diameters of the resulting flares were measured after 30 minutes. All four subjects gave similar responses of varying intensity with flares up to 5 cm. in diameter. The reaction was negative with serum obtained before infection, and became positive 4 weeks after each infection, with a peak at 3 months. After the first infection the response had decreased markedly in 6 months, and had disappeared after one year.

Haemagglutination

The method used was essentially that described by SOULSBY and GILLES (1965). Tanned sheep red cells were coated with 1 : 750 larval extract in phosphate buffered saline (pH 7·2), formalinized, and stored at 4°C. for up to 6 months. Tanned and formalinized, but uncoated, cells were used to pre-absorb sera before testing, one drop of packed cells being incubated with 5 drops of a 1 : 20 dilution of serum at room temperature for one hour. Coated cells were added to serial 2-fold dilutions of the pre-absorbed serum in borate-succinate-albumin (1%) buffer at pH 7·2, using Perspex trays and Takatsy loops, and were left overnight before reading.

Most unabsorbed sera of adult Nigerians agglutinated tanned uncoated sheep cells, but a single pre-absorption was effective in preventing such agglutination in all but about 5%. Pre-treatment of serum with 0·2% 2-mercaptoethanol prevented agglutination of uncoated cells, in agreement with the findings of ADENIYI-JONES (1967) that the antibody involved is IgM, but reduced titres with coated cells by at most one dilution. IgG was separated from one high titre serum on a diethylaminoethyl cellulose column, and was shown to be pure by immunoelectrophoresis. At a concentration of 2 g. per 100 ml. it gave a titre of 1 : 1280, against 1 : 2560 for the whole serum.

160

All 100 English sera tested, and the 16 Nigerian children with ascariasis, were negative. Sera from Nigerian adults have shown titres varying from negative to 1 : 2560 (BALL, 1966). In the experimental infection the reaction remained negative for 2 years, and then became positive to a titre of 1 : 40 (Figure 2). No close relationship has been found between haemagglutination and FAT or CFT titres, or between haemagglutination titres and egg counts (Figure 4), except that the reaction was usually negative when counts were very low.

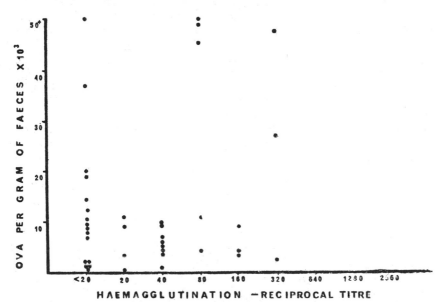

FIG. 4. The relationship between the concentration of ova in the stools and the haemagglutination reaction to *N. americanus* in 40 infected adults

Discussion

Several antibodies are produced in response to infection with *Necator americanus*, as to other nematodes with a similar life history. None of these antibodies is necessarily associated with any protective immunity. Indeed we have failed so far to produce any protective immunity by repeated experimental infection (Figure 1). The findings obtained with the FAT suggest that positive reactions are of 2 kinds, one the consequence of infection and the other spontaneous, and that there may be a qualitative difference between them.

The precipitate seen on larvae of *N. americanus* incubated in the serum of infected subjects appears to be the result of the reaction of antibody with soluble antigen on their surfaces, and to be the site of fluorescence in the FAT. In this it resembles the visible precipitate which forms around the larvae of *Ascaris* in immune serum (TAFFS and VOLLER, 1963). A high titre clearly did not prevent the normal maturation of larvae.

The apparently spontaneously occurring antibodies which we observed may be the same as those previously demonstrated in human serum to the cuticles of larvae which are normally parasitic in animals (HOGARTH-SCOTT, 1968) or are free-living (COOMBS, POUT and SOULSBY, 1965). Larvae of *N. americanus* were conspicuously less often spontaneously positive in our experiments than those of the other species tested, and then only to a low titre. This finding may be an example of the loss of antigenicity which was proposed by DINEEN (1963) to accompany adaptation to a specific host, so enabling more worms to colonize the host before the development of protective immunity. The spontaneous antigenicity of *A. duodenale* is admittedly difficult to explain on this basis, and to do so one would have to assume that its antigenicity and its greater pathogenicity are both evidence of incomplete adaptation.

The serological tests described are disappointing as prospective diagnostic tools. The FAT or the less sensitive CFT may have value in the detection of recent larval challenge, particularly if they are repeated after an interval. Their titres fall too rapidly in the absence of reinfection for them to have any place in the diagnosis of established or past infection.

SOULSBY and GILLES (1965) suggested, from the age at which they appeared in children, that haemagglutinating antibodies were produced after a delay of years, and we found this to be the case in an experimental infection. The evidence does not show whether their production is stimulated by repeated challenge with larvae or by persistence of worms in the gut, though the absence of any close relationship between titre and worm load perhaps makes the latter unlikely. Diagnostically, the haemagglutination reaction is unhelpful, because of its delayed appearance and because it is commonly negative in light infections.

REFERENCES

ADENIYI-JONES, C. (1967). *Lancet*, **1**, 188.
BALL, P. A. J. (1966). *The Pathology of parasitic diseases: Fourth symposium of the British Society for Parasitology*. Oxford: Blackwell, pp. 41-48.
COOMBS, R. R. A., POUT, D. D. & SOULSBY, E. J. L. (1965). *Exp. Parasit.*, **16**, 311.
DINEEN, J. K. (1963). *Nature, Lond.*, **197**, 268.
GILLES, H. M., WATSON-WILLIAMS, E. J. & BALL, P. A. J. (1964). *Quart. J. Med.*, **33**, 1.
HOGARTH-SCOTT, R. S. (1968). *Parasitology*, **58**, 221.
SOULSBY, E. J. L. & GILLES, H. M. (1965). *J. Parasit.*, **51**, 2 (Section 2), 39.
STOLL, N. R. & HAUSHEER, W. C. (1926). *Am. J. Hyg.*, **6**, 80.
TARR, L. F. & VOLLER, A. (1963). *Trans. R. Soc. trop. Med. Hyg.*, **57**, 353.

A Method of Labelling Living Parasitic Nematodes with Carbon 14

by

P. A. J. Ball and Ann Bartlett

The free living larvae of parasitic nematodes have been labelled with carbon 14. A labelled protein hydrolysate (specific activity 2 mc. per mg.) is used, derived from cultures of *Chlorella* (The Radiochemical Centre, Amersham). 5 µc., contained in 1 ml., is added to about 2000 ova on filter paper. The larvae take up the activity diffusely, and retain it during their parasitic development and as adults, possibly because most body cells of nematodes grow rather than divide. The labelled worms, either whole or in tissue sections, are identified by contact autoradiography with Kodirex film and an exposure of 7 days or less.

In the case of *Nematospiroides dubius* the irradiation involved does not affect the number of larvae which reach the third stage, the proportion of a dose which matures in the mouse, or the survival of adult worms 40 days after infection. Labelled larvae of *Nippostrongylus brasiliensis* have been followed to maturity. The larvae of *Necator americanus* take up activity well, but have not yet been followed past the infective stage.

Host-Nematode Interactions and
Immune Response to Nematode Infection

Bladder Tumorigenesis

D. R. STOLTZ

I. K. BARKER

Bladder tumors have been induced in rats by a number of industrial chemicals and by cyclamate. There is, however, one factor implicated in bladder tumorigenesis in rats which is frequently ignored. We refer to the possible parasitism among the rats used in these studies by the bladder nematode *Trichosomoides crassicauda* and to the possible potentiative effect of this parasite on the chemical induction of bladder tumors.

The adult *T. crassicauda* resides in the bladder, renal pelvis, and ureter, usually stimulating little tissue reaction. Parasitism with this nematode is endemic in wild rats (*1*). In 1964 Chapman (*2*) examined rats from 19 commercial American sources and found that approximately one-half of the groups contained at least one parasitized animal. The incidence of infection in these groups varied from 4 to 91 percent (11 to 24 rats per group). Although a few cesarian-delivered rats were free of parasites, one cannot dismiss the possibility of transplacental infection or subsequent postnatal infection. Dissemination of infection occurs readily since embryonated eggs that are passed in the urine are immediately infective. Contaminated cages, unless washed thoroughly with sufficiently hot water, can cause a large proportion of a colony to become infected.

A report by Chapman (3) indicates that infection with *T. crassicauda* may increase the incidence of bladder tumors in rats fed the well-known bladder carcinogen 2-acetylaminofluorene. A somewhat dated controversy as to whether *T. crassicauda* is associated, in the absence of exogenous carcinogens, with bladder tumors (3, p. 154) need not be invoked.

Thus it would appear desirable that any investigator encountering bladder tumors in rats make a thorough search for this parasite. Methods are available for eliminating the infection and for maintaining a clean rat colony (4).

References

1. L. J. Thomas, *J. Parasitol.* **10**, 105 (1924); V. S. Smith, *ibid.* **32**, 142 (1946).
2. W. H. Chapman, *Invest. Urol.* **2**, 52 (1964).
3. ———, *ibid.* **7**, 154 (1969).
4. J. F. Bone and J. R. Harr, *Lab. Animal Care* **17**, 321 (1967); D. V. Wahl and W. H. Chapman, *ibid.*, p. 386.

The effect of a
low protein diet and a glucose and filter paper diet on the
course of infection of *Nippostrongylus brasiliensis*

By KAREN R. CLARKE

INTRODUCTION

In general, malnutrition results in increased susceptibility and decreased acquired resistance to parasites; in particular this has been confirmed for *Nippostrongylus brasiliensis* in the white rat for generally deficient diets (Chandler, 1932), a milk diet (Porter, 1935*b*), deficiencies of vitamin A (Spindler, 1933; Kaneko, 1939; Watt, Golden, Olason & Mladinich, 1943; Riley, 1943), vitamin C (Matsumori, 1941) and vitamins B1 and B2 (Watt, 1944). However, deficiencies of vitamin B12 and pteroglutamic acid apparently had no effect on *N. brasiliensis* (Maldonado & Asenjo, 1953). The effect of variation in the protein content of the diet has been investigated; a high level (22 %) resulted in better growth of the rats, while the growth of the worms was retarded and their numbers reduced (Miyasaka, 1941). A diet with only 9 % protein resulted in an increased number of worms in primary infections and reduced acquired resistance (Donaldson & Otto, 1946). The effect of the quality of protein as distinct from quantity revealed that animals fed plant protein (arachin) instead of animal protein (casein) developed a lower degree of acquired resistance to *N. brasiliensis* (Barakat, 1948). The Agricultural Research Council low protein diet used in the present experiments was found to reduce acquired resistance (Wells, 1962) and increase susceptibility (Clarke, 1967). There is general agreement by the above authors that the deficient diets cause changes in the host's physiology rather than affecting the parasites directly. It has been suggested that interference with the development and maintenance of immunity occurs (Watt *et al.* 1943; Riley, 1943; Watt, 1944; Donaldson & Otto, 1946; Barakat, 1948; Wells, 1962). There is certainly a considerable amount of indirect evidence to support this idea, but while this could well account for animals being unable to throw off an infection, become resistant to repeated infections or lose any resistance that was developed while on an adequate diet, it does not account for an increase in susceptibility. This is the initial response to the parasite before antibodies have been developed and represents the condition of the animal before infection. Antibodies have not so far been detected earlier than 3 weeks after infection (Sulzer, 1964), whereas effects of protein deficiency can be seen long before this. It has been suggested that deficient diets might produce their effects by stressing the animal with a resulting increase in corticosteroids (Read, 1958). The fact that injecting cortisone and, to a lesser extent, ACTH causes higher parasite burdens is well known (Cavallero & Sala 1951; Stoner & Godwin, 1953; Weinstein, 1955; Coker, 1956; Briggs, 1959; Bezubick, 1960; Cross, 1960; Parker, 1961; Oliver, 1962; Nelson, 1962). The effect of cortisone in suppressing inflammation is also established (Menkin, 1960). However, it has been shown that there is no direct connexion between malnutrition, corticosteroid level and parasitism (Clarke, 1967). Although malnutrition may result in an increase in corticosteroid level, severe dietary deficiency can cause increased susceptibility independently of this. In order to ascertain whether deficient diets affect all phases of the life cycle of tissue-migrating nematodes equally, experiments were designed to follow the course of infection of *N. brasiliensis* in rats fed on a low protein diet, and on generally deficient diet of glucose and filter paper. It was hoped that a comparision of the effects of the diets on the migratory phase with the effects on the intestinal phase would further elucidate the mechanisms by which deficient diets affect nematode infections.

MATERIALS AND METHODS

The parasite. The original stock of *N. brasiliensis* was obtained from the Wellcome laboratories of Tropical Medicine, London. They were cultured according to the method of Wilson & Dick (1964).

The host. All the rats used were from the Sheffield University Zoology Department's colony of Wistar albino rats (*Rattus norvegicus*). Male and female rats of different ages were used, the age and sex are detailed with the results for each experiment; all rats were between 13 and 15 weeks old at autopsy. The larvae (1000–2000) were injected subcutaneously at the back of the neck or on the shoulder. Precautions were taken to ensure that few larvae remained in the syringe (cf. Twohy, 1956). Rats were maintained before and after infection as described elsewhere (Clarke, 1967).

The deficient diets. The composition of the low protein (L.P.) and glucose-filter paper(G.F.P.) diets compared with the control diet (Oxoid diet 86) is shown in Table 1. Details of the preparation of the diets have been given in Clarke (1967). Both deficient diets were fed *ad lib.* in hoppers, and water was available at all times.

Autopsy for larvae in the lungs (modified from Twohy, 1956). Both lungs were removed, chopped and washed into a 100 ml flask with 10 ml of 0·9 % NaCl, 15 ml of 0·5 % HCl and 25 ml of 1·0 % pepsin (freshly made). The tissue was incubated at 37 °C in a Warburg constant temperature bath for 3 h. After incubation the pH was adjusted to 7 to prevent further digestion (with 8 ml of 1·0 % Na_2CO_3) and the flasks stored at − 1 °C for counting. The incubate was agitated and tipped into a measuring cylinder, the flask was rinsed and the total volume recorded. The incubate was again agitated, and a 10 ml sample was withdrawn; this was centrifuged and decanted to the 5 ml level. All larvae in this residual 5 ml were counted in a Petri dish under a binocular microscope, and the total number in the whole incubate was calculated.

169

Autopsy for larvae and adult worms in the intestine. The same procedure was used as previously (Clarke, 1967), but greater care was needed in checking pieces of intestine for fourth-stage larvae, since they are smaller than adults.

Measuring larvae and worms. The method for doing this has been adapted from Haley & Parker (1961). Saline containing larvae or worms was transferred to a slide and, using a binocular microscope and dissecting needles, a parasite was isolated, straightened in a film of saline and measured by means of a micrometer scale.

Table 1. *Comparison of Oxoid* 86; *A.R.C. low protein; glucose and filter paper diets*

	Percentage in diet	
Constituents	Diet 86	Low protein (LP.)
Barley meal	25	10
Whole wheat meal	50	10
Meat and bone meal	6	—
White fish meal	7	—
Dried grass	5	—
Dried yeast	5	—
Salt	1	—
Tapioca meal	—	30
Molasses	—	10
Potatoes	—	40
Minerals	—	2 (extra)
Cod liver oil	1	1 (extra)
Protein: (Carbohydrate + fat)	1:2·8	1:25·7
Total protein percentage	20	4

Glucose and filter-paper diet (G.F.P.)

Analar glucose	80 %
Whatman number 1 filter paper	20 %

Plan of investigation. The effect of the deficient diets on the number of larvae reaching the lungs was studied by maintaining the rats on the L.P. diet for 51 days and on the G.F.P. diet for 23 days. Autopsies were at 18, 22 and 26 h after infection. The investigation was not continued beyond this point since, although larvae will still be arriving from the skin, others leave the lungs for the intestine between 37 and 41·5 h after infection (cf. Twohy, 1956) and hence confuse the analysis. A check on the total number of larvae passing through the lungs was obtained by an autopsy of control rats at 7 days after infection.

The effect of the two diets on the number of larvae arriving in the intestine was investigated by maintaining the rats on the L.P. diet for 56–58 days, and on the G.F.P. diet for 28–30 days; they were autopsied 48, 56, 64, 72, 80 and 96 h after infection. In a third experiment the course of infection was followed with a repeat sample at 4 days. Autopsies were also carried out 6, 10, 13 and 17 days after infection; the sex ratio of worms was determined in this experiment. From these three experiments samples of larvae and worms were obtained for measuring (Table 6). Rats on the deficient diets had larger worm burdens than controls;

these were initially lost faster. To determine whether this was due to the larger initial worm population or to the diets, the rate of loss of worms from populations of different sizes (obtained in control rats by different doses of larvae) were compared with the rate of loss of worms from rats on the L.P. and G.F.P. diets (Table 5).

RESULTS

The results showing the numbers of larvae and worms in rats on the three diets are given in Tables 2–4; and illustrated in Figs. 1–3. A substantial number of larvae were present by 18 h in the lungs of rats on all three diets; there was a particularly sharp increase in the number of larvae in control rats between 18 and 22 h

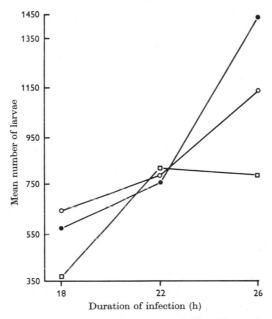

Fig. 1. Effect of L.P. and G.F.P. diets on the number of larvae in the lungs. ○, L.P. diet; ●, G.F.P. diet; □, diet 86 control.

($P < 0·01$), and there was apparently no further migration of larvae into the lungs after this (see Fig. 1). However, in L.P. and G.F.P. rats the migration continued at a rapid rate and by 26 h there was a significant difference between them and controls ($P < 0·02, P < 0·01$). The number of mature worms in control rats 7 days after infection (Table 2) was significantly greater than the number of larvae in the lungs of control rats at 26 h ($P < 0·02$), which indicates that some larvae did arrive in the lungs after 26 h.

The migration to the intestine (Table 3, Fig. 2) shows that up to 64 h the course of infection is approximately parallel in the three groups of rats; but by 72 h there was a significant difference between all three groups ($P < 0·02, P < 0·01$), with the

Table 2. *Effect of* L.P. *and* G.F.P. *diets on number of larvae in the lungs*

Group and sex	Diet	Days on diet	Number in group	Initial weight (g.)	Final weight (g.)	Percentage loss or gain in weight	Duration of infection	Larvae in lungs worms in intestine
A ♀	L.P.	51	3	128±2	95±3	−26±1·7	18 h	640±121
♀	G.F.P.	23	3	146±5	100±2	−31±1·2	18 h	572±94
♀	86 Control	—	3	115±10	155±11	−37±8·8	18 h	368±58
B ♀	L.P.	51	3	118±4	85±5	−28±2·0	22 h	799±175
♂♀	G.F.P.	23	3	171±19	113±31	−34±0·7	22 h	766±20
♀	86 Control	—	3	121±8	160±0	+33±9·0	22 h	780±8
C ♀	L.P.	51	3	103±12	74±9	−28±5·5	26 h	1130±76
♂	G.F.P.	23	3	222±8	155±9	−30±1·5	26 h	1442±119
♀	86 Control	—	3	119±7	163±10	+37±6·9	26 h	774±39
D ♀	86 Control	—	3	117±7	161±7	+38±6·6	7 days	934±18

'*t*' and *P* values: number of parasites

A L.P.–86	$t = 2·03$	$P > 0·1$		C L.P.–86	$t = 4·17$	$P < 0·02$
G.F.P.–86	$t = 1·84$	$P < 0·2$		G.F.P.–86	$t = 5·33$	$P < 0·01$
L.P.–G.F.P.	$t = 0·44$	$P < 0·7$		L.P.–G.F.P.	$t = 2·20$	$P < 0·1$
B L.P.–86	$t = 0·10$	$P > 0·9$				
G.F.P.–86	$t = 0·64$	$P < 0·6$				
L.P.–G.F.P.	$t = 0·18$	$P < 0·9$				
L.P. 18–22	$t = 0·74$	$P < 0·5$	Diet 86 18–22	$t = 7·08$	$P < 0·01$
22–26	$t = 1·73$	$P < 0·2$	22–26	$t = 0·15$	$P < 0·9$
G.F.P. 18–22	$t = 2·01$	$P < 0·1$	26–7	$t = 3·76$	$P < 0·02$
22–26	$t = 5·59$	$P < 0·01$			

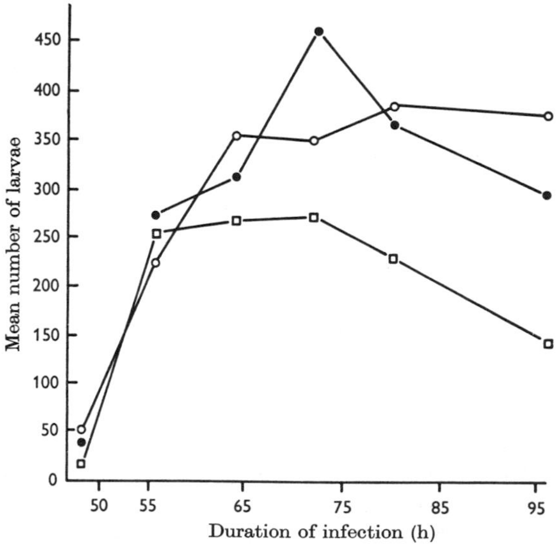

Fig. 2. Effect of L.P. and G.F.P. diets on the arrival of larvae in the intestine. ○ , L.P. diet; ●, G.F.P. diet; □, diet 86 control.

G.F.P. rats having the highest and the controls the lowest number of larvae. This situation reflects the one in the lungs 26 h after infection. By 80 h G.F.P. rats had lost some of their larvae, but rats on the deficient diets still had significantly more parasites than controls ($P > 0.01$). By 96 h both G.F.P. and control rats had lost

Table 3. *Effect of L.P. and G.F.P. diets on larval migration to the intestine*

Group	Diet	Days on diet	No. in group	Initial weight (g.)	Final weight (g.)	Percentage loss or gain in weight	Duration of infection (h)	Larvae in intestine
A	L.P.	56	3	126 ± 9	109 ± 9	-13 ± 3.3	48	48 ± 17
	G.F.P.	28	3	175 ± 19	113 ± 12	-35 ± 1.7	48	43 ± 7
	86 Control		3	121 ± 5	229 ± 19	$+89 \pm 8.0$	48	15 ± 3
B	L.P.		3	112 ± 10	92 ± 10	-18 ± 1.7	56	221 ± 14
	G.F.P.		3	182 ± 21	123 ± 14	-32 ± 2.1	56	270 ± 40
	86 Control		3	122 ± 8	213 ± 30	$+74 \pm 9.3$	56	252 ± 12
C	L.P.		3	120 ± 10	98 ± 7	-18 ± 3.4	64	355 ± 16
	G.F.P.		3	163 ± 14	103 ± 7	-37 ± 1.5	64	311 ± 21
	86 Control		3	117 ± 9	205 ± 16	$+75 \pm 1.2$	64	263 ± 30
D	L.P.	57	3	108 ± 11	86 ± 11	-21 ± 2.6	72	349 ± 14
	G.F.P.	29	3	158 ± 23	100 ± 13	-36 ± 2.6	72	460 ± 22
	86 Control		3	104 ± 13	195 ± 28	$+89 \pm 15$	72	268 ± 11
E	L.P.		3	129 ± 7	107 ± 7	-17 ± 1.5	80	382 ± 19
	G.F.P.		3	157 ± 20	97 ± 11	-38 ± 1.0	80	365 ± 3
	86 Control		3	122 ± 15	227 ± 33	$+86 \pm 12$	80	227 ± 34
F	L.P.	58	3	115 ± 14	87 ± 12	-25 ± 3.1	96	375 ± 35
	G.F.P.	30	3	185 ± 10	115 ± 13	37 ± 1.4	96	207 ± 57
	86 Control		3	101 ± 11	187 ± 32	$+82 \pm 14$	96	144 ± 22

'*t*' and *P* values: number of larvae

A	L.P.–86	$t = 1.96$	$P > 0.1$	C L.P.–86	$t = 2.69$	$P > 0.05$
	G.F.P.–86	$t = 3.91$	$P < 0.02$	G.F.P.–86	$t = 1.30$	$P < 0.3$
	L.P.–G.F.P.	$t = 0.28$	$P < 0.8$	L.P.–G.F.P.	$t = 1.65$	$P < 0.2$
B	L.P.–86	$t = 1.67$	$P < 0.2$	D L.P.–86	$t = 4.49$	$P < 0.02$
	G.F.P.–86	$t = 0.43$	$P < 0.7$	G.F.P.–86	$t = 7.88$	$P < 0.01$
	L.P.–G.F.P.	$t = 1.16$	$P > 0.3$	L.P.–G.F.P.	$t = 4.23$	$P < 0.02$
E	L.P.–86	$t = 4.04$	$P > 0.01$	F L.P.–86	$t = 5.60$	$P < 0.01$
	G.F.P.–86	$t = 4.09$	$P > 0.01$	G.F.P.–86	$t = 2.49$	$P > 0.05$
	L.P.–G.F.P.	$t = 0.90$	$P > 0.4$	L.P.–G.F.P.	$t = 1.16$	$P > 0.3$

Data from mixed groups of male and female rats.

a considerable number of parasites, while L.P. rats had not and therefore had the highest number of larvae. The repeat sample at 96 h (4 days, Table 4, Fig. 3) confirmed this. Up to 10 days after infection there was only a slight loss of worms from all three groups of rats; this trend continued in control rats up to 13 days after infection. Rats on the deficient diets lost worms after 10 days so that by 13 days there was no significant difference between the three groups ($P < 0.8$,

Table 4. *Effect of L.P. and G.F.P. diets on the course of infection in the intestine*

Group and diet	Days on diet	Number in group	Initial weight (g)	Final weight (g)	Percentage loss or gain in weight	Duration of infection (days)	Worms in intestine	♂	♀	♂:♀
A L.P.	46	3	119±2	103±4	−13±1.8	4	701±14	289±11	412±6	1:1.42±0.054
G.F.P.	15	3	168±19	136±16	−19±0.9	4	543±38	218±22	325±21	1:1.51±0.137
Control		3	109±6	181±11	+67±20.9	4	461±3	157±6	304±9	1:0.94±0.137
B L.P.	48	3	106±4	94±5	−11±2.1	6	683±19	280±11	402±14	1:1.43±0.075
G.F.P.	17	3	192±31	147±25	−24±0.7	6	618±91	228±46	390±47	1:1.78±0.199
Control		3	101±5	197±17	+96±17.0	6	507±22	203±14	304±29	1:1.52±0.223
C L.P.	52	3	121±20	103±15	−14±1.7	10	634±13	282±9	352±15	1:1.25±0.083
G.F.P.	21	3	173±31	124±18	−28±2.6	10	577±19	266±13	311±8	1:1.17±0.047
Control		2	94±4	150±10	+59±3.0	10	471±30	176±36	295±3	1:1.75±0.395
D L.P.	55	3	111±17	95±18	−14±4.0	13	413±77	178±38	234±47	1:1.35±0.272
G.F.P.	24	3	157±16	108±14	−31±2.4	13	335±83	152±39	183±46	1:1.23±0.114
Control		2	101±6	204±3	+103±13.4	13	440±26	205±39	235±14	1:1.20±0.300
E L.P.	59	2	132±16	104±14	−21±1.0	17	367±134	201±70	166±64	1:0.80±0.035
G.F.P.	28	3	205±5	134±5	−35±0.7	17	46±43	40±37	6±6	1:0.04±0.046
Control		3	89±5	149±3	+70±10.1	17	4±2	4±2	0	1:0.00

'*t*' and *P* values

Number of worms

		t	*P*
A	L.P.–86	16.35	<0.001
	G.F.P.–86	2.15	<0.1
	L.P.–G.F.P.	3.89	<0.02
B	L.P.–86	6.04	<0.01
	G.F.P.–86	1.18	>0.3
	L.P.–G.F.P.	0.70	>0.5
C	L.P.–86	4.96	<0.02
	G.F.P.–86	3.00	>0.05
	L.P.–G.F.P.	2.49	>0.05
D	L.P.–86	0.33	<0.8
	G.F.P.–86	1.20	>0.3
	L.P.–G.F.P.	0.68	>0.5
E	L.P.–86	2.71	<0.1
	G.F.P.–86	0.98	<0.4
	L.P.–G.F.P.	2.29	<0.2

		t	*P*
L.P.	6–10 days	2.11	>0.1
	10–13	2.83	<0.05
	13–17	0.29	<0.8
G.F.P.	6–10 days	0.44	<0.7
	10–13	2.83	<0.05
	13–17	3.08	<0.05
Control diet 86	6–10 days	0.96	>0.4
	10–13	0.78	<0.6
	13–17	16.71	<0.001

Ratio—male:female

		t	*P*
A	L.P.–86	3.53	<0.05
	G.F.P.–86	2.22	<0.1
	L.P.–G.F.P.	0.63	>0.5
B	L.P.–86	0.42	>0.6
	G.F.P.–86	0.91	>0.4
	L.P.–G.F.P.	1.64	>0.2
C	L.P.–86	3.33	<0.05
	G.F.P.–86	4.34	<0.05
	L.P.–G.F.P.	0.83	>0.5
D	L.P.–86	0.37	>0.7
	G.F.P.–86	0.09	>0.9
	L.P.–G.F.P.	0.40	>0.7
E	L.P.–86	22.85	<0.001
	G.F.P.–86	0.86	<0.5
	L.P.–G.F.P.	13.14	<0.001

The data are for 21 male and 22 female rats in mixed groups. Rats were started on the L.P. diet at 7 weeks, and on the G.F.P. diet at 10 weeks.

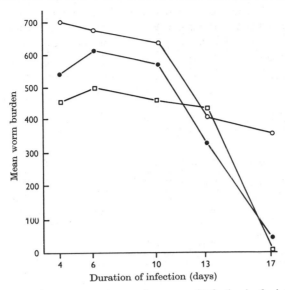

Fig. 3. Effect of L.P. and G.F.P. diets on the course of infection in the intestine.
○, L.P. diet; ●, G.F.P. diet; □, diet 86 control.

Table 5. *Effect of dose size on course of infection in intestine*

Group	Number in group	Initial weight (g.)	Final weight (g.)	Dose	Duration of infection (days)	Worms in intestine	Percentage in intestine
A	3	189 ± 26	192 ± 28	700	7	337 ± 13	100 ± 4·1
B	3	193 ± 23	199 ± 20	2000	7	868 ± 19	100 ± 2·0
C	3	206 ± 34	203 ± 29	3000	7	1108 ± 36	100 ± 3·1
D	3	170 ± 25	169 ± 25	700	10	306 ± 35	91 ± 10·5
E	3	178 ± 28	181 ± 30	2000	10	851 ± 19	98 ± 2·3
F	3	164 ± 24	174 ± 26	3000	10	950 ± 36	86 ± 3·2
G	3	203 ± 22	214 ± 26	700	13	317 ± 21	94 ± 6·4
H	3	212 ± 26	219 ± 29	2000	13	777 ± 57	89 ± 6·7
I	3	189 ± 26	198 ± 30	3000	13	978 ± 88	88 ± 8·0
J	3	179 ± 27	191 ± 33	700	17	121 ± 47	36 ± 13·8
K	3	164 ± 20	175 ± 25	2000	17	304 ± 145	35 ± 16·6
L	3	163 ± 24	174 ± 31	3000	17	335 ± 117	30 ± 15·8

'*t*' and *P* values:

Dose of 700

A–D $t = 0.83$ $P < 0.5$
D–G $t = 0.27$ $P > 0.8$
G–J $t = 3.83$ $P < 0.02$

Dose of 2000

B–E $t = 0.63$ $P < 0.6$
E–H $t = 1.18$ $P > 0.3$
H–K $t = 3.02$ $P > 0.02$

Dose of 3000

C–F $t = 3.10$ $P > 0.02$
F–I $t = 0.29$ $P < 0.8$
I–L $t = 3.25$ $P > 0.02$

The data are for 18 male and 18 female rats, 10–12 weeks old.

$P > 0.3, P > 0.5$). The loss of worms continued in G.F.P. rats, so that both G.F.P. and control rats had lost the bulk of their infections by 17 days; however, at this stage L.P. rats still retained about 50 % of the worm burden they had on the fourth day.

Table 6. *Effect of L.P. and G.F.P. diets on size of larvae, immature and adult worms*

Time after infection (h)	Stage	L.P. (mm)	G.F.P. (mm)	86 control (mm)	Table number
0	L 3	—	—	0.603 ± 0.0087	—
18	L 3	0.682 ± 0.0097	0.680 ± 0.0059	0.689 ± 0.0091	2
22	L 3	0.741 ± 0.0115	0.746 ± 0.0105	0.727 ± 0.0128	2
26	L 3	0.768 ± 0.0144	0.762 ± 0.0140	0.762 ± 0.0117	2
48	L 4	1.310 ± 0.0223	1.113 ± 0.0243	1.270 ± 0.0210	3
64	L 4	1.748 ± 0.0275	1.820 ± 0.0303	1.772 ± 0.0340	3
80	L 4 ♂	1.972 ± 0.0302	2.038 ± 0.0314	2.026 ± 0.0470	3
	L 4 ♀	2.390 ± 0.0486	2.373 ± 0.0374	2.314 ± 0.0577	3
96	L 4 ♂	2.439 ± 0.0620	2.325 ± 0.0670	2.706 ± 0.0534	4
	L 4 ♀	2.825 ± 0.566	2.732 ± 0.0521	2.771 ± 0.0696	4
144	Adult ♂	3.842 ± 0.0306	3.804 ± 0.0417	3.786 ± 0.0447	4
	Adult ♀	4.801 ± 0.0612	4.942 ± 0.0641	4.783 ± 0.0537	4

't' and P values

18 h			80 h male		
L.P.–86	$t = 0.52$	$P > 0.6$	L.P.–86	$t = 0.96$	$P < 0.4$
G.F.P.–86	$t = 0.82$	$P > 0.4$	G.F.P.–86	$t = 0.21$	$P > 0.8$
L.P.–G.F.P.	$t = 0.17$	$P < 0.9$	L.P.–G.F.P.	$t = 1.51$	$P < 0.2$

22 h			80 h Female		
L.P.–86	$t = 0.81$	$P > 0.4$	L.P.–86	$t = 1.00$	$P > 0.3$
G.F.P.–86	$t = 1.15$	$P < 0.3$	G.F.P.–86	$t = 0.85$	$P < 0.4$
L.P.–G.F.P.	$t = 0.32$	$P > 0.7$	L.P.–G.F.P.	$t = 0.27$	$P < 0.8$

26 h			96 h Male		
L.P.–86	$t = 0.32$	$P > 0.7$	L.P.–86	$t = 3.26$	$P < 0.001$
G.F.P.–86	$t = 0$	$P > 0.9$	G.F.P.–86	$t = 4.44$	$P < 0.001$
L.P.–G.F.P.	$t = 0.29$	$P < 0.8$	L.P.–G.F.P.	$t = 1.24$	$P < 0.3$

48 h			96 h Female		
L.P.–86	$t = 1.30$	$P < 0.2$	L.P.–86	$t = 0.60$	$P < 0.6$
G.F.P.–86	$t = 4.88$	$P < 0.001$	G.F.P.–86	$t = 0.44$	$P < 0.7$
L.P.–G.F.P.	$t = 5.97$	$P < 0.001$	L.P.–G.F.P.	$t = 1.23$	$P < 0.3$

64 h			144 h Male		
L.P.–86	$t = 0.54$	$P < 0.6$	L.P.–86	$t = 1.04$	$P > 0.3$
G.F.P.–86	$t = 1.05$	$P < 0.3$	G.F.P.–86	$t = 0.29$	$P < 0.8$
L.P.–G.F.P.	$t = 1.75$	$P < 0.1$	L.P.–G.F.P.	$t = 0.73$	$P < 0.5$

144 h Female		
L.P.–86	$t = 0.22$	$P > 0.8$
G.F.P.–86	$t = 1.90$	$P < 0.1$
L.P.–G.F.P.	$t = 1.59$	$P < 0.2$

This shows the size of parasites from three experiments. The rest of the relevant data can be found by referring to the appropriate table. Numbers measured varied from 15 to 30 according to the degree of variation exhibited.

In order to test whether the earlier loss of worms from rats on the deficient diets, which had also been noted in earlier experiments (Clarke, 1967) could have been due to the initially larger worm burden, the effect of different sized worm burdens on the course of infection in control rats was examined. The results (Table 5) show a significant loss of worms by all three groups between 13 and 17 days after infection. There was also a significant loss of worms by the rats infected with 3000 larvae between 7 and 10 days after infection. However, since the groups infected with 700 and 2000 larvae produced a similar range in number of worms to those found in rats on the L.P. and G.F.P. diets (Tables 3, 4) and did not lose worms in significant numbers before 13 days, it is concluded that the larger worm burdens were not responsible for the earlier loss of worms in L.P. and G.F.P. rats.

No consistent pattern of difference emerged between the ratio of male to female worms in animals on the three diets, although in the 4-day infection there were significantly fewer female worms in L.P. rats than in controls ($P < 0.05$). This also occurred at 10 days when both L.P. and G.F.P. rats had fewer female worms than controls ($P < 0.05$). By 17 days the L.P. rats had significantly more female worms than G.F.P. or control rats ($P < 0.001$) which had lost the bulk of their infections.

The lengths of larvae and adult worms from rats on the three diets are shown in Table 6, and apart from two occasions there were no significant differences. The exceptions were the 48 h infection, in which the larvae from G.F.P. rats were significantly smaller than those from L.P. or control animals; and the 96 h infection where worms from both L.P. and G.F.P. rats were significantly smaller than controls.

DISCUSSION

In rats on the deficient diets significantly larger numbers of larvae reached the lungs than in control rats, though the migration proceeded at the same rate in all groups of animals. This confirms the general impression given by the literature that deficient diets result in increased susceptibility to nematode infection. The difference between rats on the deficient diets and controls was accounted for by the migration continuing for longer in the former animals. It has been suggested that a more rapid migration of larvae takes place in rats on deficient diets (Donaldson & Otto, 1946), but this did not occur in the present experiments. The above authors also suggested that the parasites might develop more rapidly in animals on deficient diets, but measurements of larvae and worms have shown that this is not the case. It is thought unlikely (Clarke, 1967) that the increase in susceptibility, which is shown by the increased number of larvae completing the migration to the intestine of rats fed the deficient diets, is due to a failure in the development of antibodies. This is supported by the fact that in a case of increased susceptibility to fatal bacterial infection in rats mildly deficient in protein, no reduction in antibody titre occurred (Miles, 1951). It was therefore suggested that some other factor might be involved. This could be a reduced cellular reaction in the early stages of infection. Cells of the reticulo-endothelial system are closely concerned with this and it is perhaps significant that a 2 % protein diet causes atrophy of the bone marrow, liver and spleen, and a significant reduction in lymphocytes and leucocytes (Asirvadham, 1948).

In rats on the G.F.P. diet, which had more larvae in the lungs and significantly more in the intestine than L.P. rats, there was an increased corticosteroid level (Clarke, 1967) and it appears that the combined effect of the deficient diet plus the raised corticosteroid level resulted in even more larvae completing the migration than where only one of these factors was operating.

The lung-intestinal migration proceeded at the same rate in animals on all three diets, as had the skin-lung migration; but in rats on the deficient diets the migration continued longer so that significantly more larvae were found in the intestine. The maximum number of larvae was found in the intestine between 72 and 80 h after infection; this applied to all three groups of animals. It was also noticed that although the G.F.P. rats had significantly more larvae in the intestine than L.P. rats at 72 h, by 96 h this situation had been reversed. The pattern of the lung-intestinal migration again indicates that there is not a more rapid migration of parasites in the animals on the deficient diets. Control and G.F.P. rats lost a considerable number of larvae between 72 h and 96 h but this did not occur in L.P. rats. Since this approximately coincides with the time of the final moult (Yokogawa, 1922), it is possible that the larvae loosen their hold on the wall of the intestine and during the moulting process are swept out by the gut contents. The moulting process has been observed *in vitro*; a longitudinal rupture occurs in the old cuticle at the base of the oesophagus and the larva works its way sideways through this opening, the anterior region of the old cuticle becoming invaginated. If the moult occurs in a similar manner *in vivo* it appears likely that the worms would become less intimately associated with the mucosa of the intestine. It is further possible that intestinal mobility may be affected by the diets, since it has been shown that *Aspiculuris tetraptera*, *Syphacia obvelata* and *Hymenolepis nana* infections in mice can be affected by intestinal mobility (De Witt & Weinstein, 1964). As the faeces of L.P. rats were rather softer than normal, friction and intestinal mobility may have been reduced, thereby accounting for these rats retaining all the larvae that reached the intestine. It is hoped to carry this investigation further.

More larvae were found in the intestine of control rats autopsied at 7 days than in the lungs at 26 h. A similar occurrence has been noted by Twohy (1956), who recovered from the skin and lungs only 60 % of the infective dose. It was assumed that a loss of 40 % occurred during the processing of the tissues and counting of the larvae. However, it is quite possible that in the present experiments, and those of Twohy, the missing larvae did not go to the lungs or only passed through very rapidly. Experiments in which larvae were injected into the portal vein and developed in the liver indicate that the lungs are not the only tissue with the physiological requirements for the early growth of *N. brasiliensis* (Twohy, 1955).

Between 10 and 13 days after infection L.P. and G.F.P. rats lost worms more rapidly than control rats. This indicates that although the deficient diets favour the larval phase of the parasite, conditions in the intestine are not favourable to the adult worms. This is further indicated by the fact that on the occasions when there was a significant difference in size between larvae or adult worms from rats on the deficient diets and controls, parasites from the former were smaller. Since this occurred at 48 and 96 h (male worms only) a delay in growth after the moult or a delay in the moult may have taken place. It has been shown in the present experiments that the loss of worms between 10 and 13 days was not due to the larger initial worm burden in the rats on the deficient diets. It is therefore possible that the diets were affecting the worms themselves possibly by altering the composition of the blood or tissues on which the worms feed (Porter, 1935b; Taliaferro & Sarles, 1939; Rogers & Lazarus, 1949; Haley & Parker, 1961). In connexion with this, information on the age at which sexual maturity of the worms occurs, level of egg production and viability of eggs of parasites from animals fed deficient diets would be interesting. However, it has been shown that no significant difference occurred in the level of plasma protein due to qualitative protein deficiency; though lower levels of haemoglobin were observed (Barakat, 1948). In the absence of further information on circulating levels of nutrients in animals fed on deficient diets it is suggested that this loss of worms in the intestine may be due to the lack, or low level, of some common dietary factor in animals on both deficient diets. In connexion with this it was noticed that food consumption dropped in rats on the L.P. diet towards the end of an experiment. Experiments on the *in vitro* cultivation of the parasitic stages of *N. brasiliensis* indicated that some dietary factor was missing, as only stunted adults developed (Weinstein & Jones, 1956a). Later it was found that the loss of water soluble vitamins was one of the most important factors limiting growth (Weinstein & Jones, 1956b, 1957). In addition, it has been shown that the rate of absorption of glucose from the jejunum is depressed in rats with *N. brasiliensis* and protein digestion is also affected (Symons, 1960a, b); further, there is decreased enzyme activity (Symons & Fairbairn, 1962, 1963). While this impaired digestion might not affect well-fed rats and their parasites, it could accentuate the low level of essential nutrients in rats fed deficient diets following a mature infection in the intestine.

Further evidence of the deficient diets having unfavourable effects on adult worms is indicated by significantly fewer female worms being present in L.P. and G.F.P. rats. It appears that fewer female worms indicates an unfavourable environment, since the normal sex ratio of one male to two females (Africa, 1931; Porter, 1935a; Haley, 1962) is reversed in an initial contact with an abnormal host, the hamster (Haley, 1966). In the present experiments, L.P. rats had significantly more female worms by 17 days than G.F.P. and control rats. This was due to the loss of the bulk of their infections by the latter groups; female worms are lost first the last few worms of an infection being mostly males (Africa, 1931; Porter, 1935a; Haley, 1962).

Both control rats and those on the G.F.P. diet eliminated most of their worms by 17 days after infection, but L.P. rats still retained about 50 % of their 4-day worm burden. Prolonged infection due to deficient diets has been noted in several cases (Chandler, 1932; Riley, 1943; Donaldson & Otto, 1946; Wells, 1962), and it is thought on good, but indirect, evidence that the development and maintenance of antibody levels may be interfered with (Riley, 1943; Watt et al. 1943; Watt, 1944; Donaldson & Otto, 1946; Barakat, 1948). Attention has been drawn to the fact that the induction of antibody formation takes place in bacterial infections in 24–48 h, and for about 16 days the titre rises to a maximum (Miles & Wilson, 1946). This is of interest, as *N. brasiliensis* infections are normally terminated towards the

179

end of the second or third week of infection. However, to date, it has not been possible to detect antibodies to primary *N. brasiliensis* infections earlier than 3 weeks after infection (Sulzer, 1964); therefore direct investigation of the antibody level is not yet possible. It has been found that rats on the L.P. diet have a reduced capacity for mucin secretion, and it was suggested that if mucin acts as a vehicle for antibodies this could be a reason why these rats are unable to throw off an infection (Wells, 1963).

The effect of diet on the ability to produce antibodies has been studied for some host–parasite relationships. In cases of minor protein deficiency normal agglutinin responses occurred in rats infected with *Salmonella typhimurium* and *Corynebacterium murium*, though animals showed increased susceptibility (Miles, 1951). In several cases of severe dietary deficiency a reduction in antibody titre was observed, for example in rabbits on a low protein diet (Cannon, 1942); in protein depleted rats (Wissler, Woolridge & Steffee, 1946), and in chickens on a vitamin A deficient diet (Leutskaya, 1964).

There is, therefore, evidence that in severe dietary deficiency, particularly of protein, antibody formation may be impaired and it is suggested that rats on the L.P. diet come into this category in the present experiments.

Rats on the G.F.P. diet, however, were able to terminate their infections almost as soon as control rats, suggesting that their immunological response was not significantly affected. Work on the immunological status of L.P. and G.F.P. rats is in progress.

Therefore, in conclusion, it appears that deficient diets favour the migratory phase of *N. brasiliensis*, but do not necessarily provide more favourable conditions for the adult worms. Moreover, it is suggested that the deficient diets do not result in increased susceptibility and decreased resistance by affecting the same combination of mechanisms.

SUMMARY

Rats were fed on two deficient diets and were infected with *N. brasiliensis* larvae. They were autopsied at times ranging from 18 hours to 17 days after infection. The effect of the diets on worm burden, sex ratio of the worms and size of larvae and of worms were noted. Significantly more larvae reached the lungs and intestine in rats fed on glucose and filter paper (G.F.P.) and low protein (L.P.) diets than in controls, the highest numbers coming from G.F.P. rats. Neither the migration rate nor the development of the worms was accelerated by the deficient diets. G.F.P. and control rats lost larvae soon after their arrival in the intestine. Adult worms were lost from rats on the deficient diets before they were lost in significant numbers by controls; this was not due to the initially larger worm burdens in the former rats. G.F.P. and control rats lost the bulk of their infections by 17 days; L.P. rats retained about 50 % of the worm burden they had on the fourth day. The effect of the deficient diets on the migratory and adult phase of *N. brasiliensis*, and on susceptibility and resistance is discussed.

This work was done during the tenure of a Science Research Council Student-ship. My thanks are due to Dr E. T. B. Francis for his helpful and critical super-vision, and to Professor I. Chester Jones, in whose Department the work was carried out, for the facilities he provided.

REFERENCES

AFRICA, C. M. (1931). Studies on the host relations of *Nippostrongylus muris* with special reference to age, resistance and acquired immunity. *J. Parasit.* **18**, 1–13.

ASIRVADHAM, M. (1948). The bone marrow and its leukocytic response in protein deficiency. *J. infect. Dis.* **83**, 87–100.

BARAKAT, M. R. (1948). Relationship between nutrition and parasitism in the experimental animal. Ph.D. Thesis, University of London.

BEZUBIK, B. (1960). Effect of cortisone on the susceptibility of hamsters and guinea pigs to the sheep and rabbit strain of *Strongyloides papillosus*. *J. Parasit.* **46**, Suppl. 30–1. (Abstr.)

BRIGGS, N. T. (1959). The effect of cortisone on natural and acquired responses of the white rat to infection with *Litomosoides carini*. *J. Parasit.* **45**, Suppl. 37. (Abstr.)

CANNON, P. R. (1942). Antibodies and the protein-reserves. *J. Immunol.* **44**, 107–14.

CAVALLERO, C. & SALA, G. (1951). Cortisone and infection. *Lancet* **1**, 175.

CHANDLER, A. C. (1932). Experiments on resistance of rats to superinfection with the nematode *Nippostrongylus muris*. *Am. J. Hyg.* **16**, 750–82.

CLARKE, K. R. (1967). Effect of deficient diets on corticosteroid synthesis by rat adrenals *in vitro* and its relationship to infections with *Nippostrongylus brasiliensis* (Travassos, 1914). (In the Press.)

COKER, C. M. (1956). Effects of cortisone on cellular inflammation in the musculature of mice given one infection of *Trichinella spiralis*. *J. Parasit.* **42**, 479–84.

CROSS, J. H. K. JR. (1960). The natural resistance of the white rat to *Nematospiroides dubius* and the effect of cortisone on this resistance. *J. Parasit.* **46**, 175–85.

DE WITT, W. B. & WEINSTEIN, P. P. (1964). Elimination of intestinal helminths of mice by feeding purified diets. *J. Parasit.* **50**, 429–34.

DONALDSON, A. W. & OTTO, G. F. (1946). Effects of protein-deficient diets on immunity to a nematode (*Nippostrongylus muris*) infection. *Am. J. Hyg.* **44**, 384–400.

HALEY, A. J. (1962). Biology of the rat nematode *Nippostrongylus brasiliensis* (Travassos, 1914). II. Preparasitic stages and development in the laboratory rat. *J. Parasit.* **48**, 13–23.

HALEY, A. J. (1966). Biology of the rat nematode *Nippostrongylus brasiliensis* (Travassos, 1914). III. Characteristics of *N. brasiliensis* after 30–120 serial passages in the Syrian hamster. *J. Parasit.* **52**, 98–109.

HALEY, A. J. & PARKER, J. C. (1961). Size of adult *Nippostrongylus brasiliensis* from light and heavy infections in laboratory rats. *J. Parasit.* **47**, 461.

KANEKO, T. (1939). Infection experiments of *Nippostrongylus muris* in the rat, in avita minosis. *Keio Igaku* **19**, 199–212.

LEUTSKAYA, Z. K. (1964). The antibody level in vitamin A deficiency in chickens immunized with *Ascaridia galli* antigen. *Dokl. Akad. Nauk SSSR.* **159**, 938–40.

MALDONADO, J. F. & ASENJO, C. F. (1953). The role of pteroglutamic acid and vitamin B 12 on the development of *Nippostrongylus muris* in the rat. *Expl Parasit.* **2**, 374–9.

MATSUMORI, M. (1941). Infection experiments of *Nippostrongylus muris* in the rat fed with diets destitute of and added with vitamin C. *Keio Igaku* **21**, 1261–73.

MENKIN, V. (1960). Biochemical mechanisms in inflammation. *Br. Med. J.* **1**, 1521–8.

MILES, A. A. & WILSON, G. S. (1946). *Topley and Wilson's Principles of Bacteriology and Immunity*, 3rd ed., vol. II. London: Ed. Arnold Ltd.

MILES, J. A. R. (1951). Observations on the course of infection with certain natural bacterial pathogens of the rat in rats on protein-deficient diets. *Br. J. exp Path.* **32**, 285–306.

MIYASAKA, K. (1941). Infection experiment of *Nippostrongylus muris* in the rat fed with a preparation of the pupa of silkworm. *Keio Igaku* **21**, 1365–75.

NELSON, W. A. (1962). Development in a sheep of resistance to the ked *Melophagus ovinus* (L.). II. Effects of adrenocorticotrophic hormone and cortisone. *Expl Parasit.* **12**, 45–51.

OLIVER, L. (1962). Studies on natural resistance of *Taenia taeniaeformis* in mice. II. The effect of cortisone. *J. Parasit.* **48**, 758–62.

181

PARKER, J. C. (1961). Effect of cortisone on the resistance of the guinea pig to infection with the rat nematode *Nippostrongylus brasiliensis*. *Expl Parasit.* **11**, 380–90.

PORTER, D. A. (1935a). A comparative study of *Nippostrongylus muris* in rats and mice. *Am. J. Hyg.* **22**, 444–66.

PORTER, D. A. (1935b). Studies on the effect of milk diet on the resistance of rats to *Nippostrongylus muris*. *Am. J. Hyg.* **22**, 467–74.

READ, C. P. (1958). Status of behavioral and physiological resistance. *Rice Inst. Pamph.* **45**, 36–54.

RILEY, E. G. (1943). The effect of various stages of vitamin A deficiency in the white rat on the resistance to *Nippostrongylus muris*. *J. infect. Dis.* **72**, 133–41.

ROGERS, W. P. & LAZARUS, M. (1949). The uptake of radioactive phosphorus from host tissues and fluids by nematode parasites. *Parasitology* **39**, 245–50.

SPINDLER, R. L. A. (1933). Relation of vitamin A to the development of resistance in rats to superinfections with an intestinal nematode *Nippostrongylus muris*. *J. Parasit.* **20**, 72.

STONER, R. D. & GODWIN, J. T. (1953). The effects of ACTH and cortisone upon susceptibility to Trichinosis in mice. *Am. J. Path.* **29**, 943–50.

SULZER, A. J. (1964). A serological study of rats and rabbits exposed to *Nippostrongylus brasiliensis* (Travassos, 1914). *Diss. Abstr.* **24**, 3938–9.

SYMONS, L. E. A. (1960a). Pathology of infestation of the rat with *Nippostrongylus muris* (Yokogawa). IV. The absorption of glucose and histidine. *Aust. J. biol. Sci.* **13**, 180–7.

SYMONS, L. E. A. (1960b). Pathology of infestation of the rat with *Nippostrongylus muris* (Yokogawa). V. Protein digestion. *Aust. J. biol. Sci.* **13**, 579–83.

SYMONS, L. E. A. & FAIRBAIRN, D. (1962). Pathology, absorption, transport and activity of digestive enzymes in the rat jejunum parasitised by the nematode *Nippostrongylus brasiliensis*. *Fedn Proc. Fedn Am. Socs. exp. Biol.* **21**, 913–18.

SYMONS, L. E. A. & FAIRBAIRN, D. (1963). Biochemical pathology of the rat jejunum parasitised by the nematode *Nippostrongylus brasiliensis*. *Expl Parasit.* **13**, 284–304.

TALIAFERRO, W. H. & SARLES, M. P. (1939). The cellular reactions in the skin, lungs and intestine of normal, and immune rats after infection with *Nippostrongylus muris*. *J. infect. Dis.* **64**, 157–92.

TWOHY, D. W. (1955). The role of the skin and the lungs in the development of *Nippostrongylus muris*. *J. Parasit.* **41**, Suppl. 44. (Abstr.).

TWOHY, D. W. (1956). The early migration and growth of *Nippostrongylus muris* in the rat. *Am. J. Hyg.* **63**, 165–85.

WATT, J. Y. C. (1944). The influence of vitamin B1 (Thiamine) and B2 (Riboflavin) upon resistance of rats to infection with *Nippostrongylus muris*. *Am. J. Hyg.* **39**, 145–51.

WATT, J. Y. C., GOLDEN, W. R. C., OLASON, F. & MLADINICH, G. (1943). The relationship of vitamin A to resistance to *Nippostrongylus muris*. *Science* **97**, 381–2.

WEINSTEIN, P. P. (1955). The effect of cortisone on the immune response of the white rat to *Nippostrongylus muris*. *Am. J. trop. Med. Hyg.* **4**, 61–74.

WEINSTEIN, P. P. & JONES, M. F. (1956a). The in vitro cultivation of *Nippostrongylus muris* to the adult stage. *J. Parasit.* **42**, 215–36.

WEINSTEIN, P. P. & JONES, M. F. (1956b). The effect of vitamins and protein hydrolysates on the growth in vitro of the free-living stages of *Nippostrongylus muris* under axenic conditions. *J. Parasit.* **42**, 14.

WEINSTEIN, P. P. & JONES, M. F. (1957). The development of a study on the axenic growth in vitro of *Nippostrongylus muris* to the adult stage. *Am. J. trop. Med. Hyg.* **6**, 480–4.

WELLS, P. D. (1962). Mast cell, eosinophil and histamine levels in *Nippostrongylus brasiliensis* infected rats. *Expl Parasit.* **12**, 82–101.

WELLS, P. D. (1963). Mucin-secreting cells in rats infected with *Nippostrongylus brasiliensis*. *Expl Parasit.* **14**, 15–22.

WILSON, P. A. & DICK, J. M. (1964). Culture and isolation of *Nippostrongylus brasiliensis* infective larvae for biochemical studies. *J. Helminth.* **38**, 399–404.

WISSLER, R. W., WOOLRIDGE, R. L. & STEFFEE, C. H. (1946). Influence of amino acid feeding upon antibody production in protein depleted rats. *Proc. Soc. exp. Biol. Med.* **62**, 199–203.

YOKOGAWA, S. (1922). The development of *Heligmosomum muris* Yokogawa, a nematode from the intestine of the wild rat. *Parasitology* **14**, 127–66.

The life cycle of Romanomermis sp. (Nematoda: Mermithidae) a parasite of mosquitoes

B. A. OBIAMIWE

Liverpool School of Tropical Medicine

MUSPRATT (1945) in Zambia found undescribed mermithid worms in the larvae of 7 species of the genus *Aedes*, in larvae of *Culex nebulosus* and in larvae of some anophelines, including *Anopheles gambiae*. The life cycle of this mermithid has been examined and the parasite is now thought to be a new species of the genus *Romanomermis*. The essential stages of the life cycle are:

1. *Egg stage:* The egg is coffee coloured, $83 \cdot 5 \times 73 \cdot 0$ μ. The newly laid egg is at a late cleavage stage and a motile larva is formed within 10 days. As has been observed in other mermithids (CHRISTIE, 1936; POINAR, 1968), the first worm moult probably takes place inside the egg.

2. *Pre-parasitic juvenile stage:* This begins when the mermithid hatches from the egg. It is about 800μ long with a fine tail and a blunt head, and possesses an odontostyle. The stage is short-lived.

3. *Parasitic juvenile stage:* The penetration of the mosquito larva was demonstrated. The pre-parasitic juvenile bores through the cuticle of the mosquito larva and enters the haemocoele. Melanin is deposited around the point of penetration. Mermithid parasites stained in iodine, after 4 days of infection, show the characteristic 0 pairs of stichocytes and the long tubular oesophagus. A sausage stage was demonstrated. The second mermithid moult occurs about 8 days after infection. Preparations of older infections with worms thrown into several loops in the haemocoele were shown. Mosquitoes with infections over 8 days are pale so that the parasites can be seen through the body wall. The parasitic mermithids in both larval and adult *Culex p. molestus* emerge after about 14 days of infection at 25°-27°C. Up to 7 female worms have emerged from heavily infected *Culex p. molestus*.

4. *Post-parasitic juvenile stage:* Emergence of worms causes the death of the mosquito hosts. Male and female worms can be recognized. The female worm possesses a vulval primordium near the middle of the body. In the male the distance between the end of the trophosome and the end of the body is about $171 \cdot 6$ μ. In the female this distance is about $50 \cdot 3$ μ. The male worm is about $7 \cdot 3$ mm. long, the female about twice this length. The post-parasitic worms burrow into soil under water. The 3rd and 4th moults, which occur simultaneously, were demonstrated.

5. *Adult stage:* The adult mermithid loses the short pointed tail in the final moults. Male and female worms copulate and only the fertilized female lays eggs. Egg laying continues until death when the female's food reserve in the trophosome is depleted.

REFERENCES

CHRISTIE, J. R. (1936). *J. agric. Res.*, **52**, 161.
MUSPRATT, J. (1945). *J. ent. Soc. sth. Afr.*, **8**, 13.
POINAR, G. O. JR. (1968). *Proc. helminth. Soc. Wash.*, **35**, 161.

Raymond Cypess

Artificial Production of Acquired Immunity in Mice by Footpad Injections of a Crude Larval Extract of *Nematospiroides dubius*

It has been possible to produce an effective level of immunity in experimental hosts by injections of antigens prepared from a variety of larval or adult nematodes. In the case of *N. dubius*, Van Zandt (1962, J. Parasit. **48:** 249–251) used an antigen prepared by extracting ground larvae in saline for 7 days. Mice given six subcutaneous injections of the supernatant, obtained after centrifugation of the ground larvae–saline mixture, showed a significantly lower adult worm burden than control groups 15 days after a challenging infection with 50 larvae. As expected, the percentage reduction in worm burden compared with those of the controls (14.3) was not nearly as striking as that (65.6) noted in mice given three stimulating infections before being challenged (Van Zandt, 1961, J. Elisha Mitchell Sci. Soc. **77:** 300–309).

The following experiment was performed with the hope of demonstrating a more effective preparation for the artificial production of acquired immunity against this parasite. The methods of Van Zandt (1961, J. Elisha Mitchell Sci. Soc. **77:** 300–309) were used to obtain third-stage larvae. Approximately 0.3 ml of washed, packed larvae and 5.7 ml of a commercially prepared physiologic solution (TIS-U-SOL; Travenol Laboratories, Inc.) were placed in a motor-driven Ten Broeck Tissue Grinder. The preparation of the aqueous extract antigen and determination of protein content are described elsewhere (Larsh et al., 1969, J. Parasit. **55:** 726–729). The antigen was emulsified with an equal volume of Freund's complete adjuvant prior to injection.

Thirty male mice, 12 weeks old, were divided equally into three groups. Those of Group I, the experimental mice, were injected with a total of 0.08 ml (10 μg of nematospiroides protein) of the antigen–adjuvant mixture, which was divided among the four footpads. The mice of Group II, the adjuvant control mice, were injected similarly with the same volume of Freund's complete adjuvant alone, and the Group III mice, the regular controls, were injected with 0.08 ml of saline. Fourteen days later, all three groups were inoculated as before with freshly prepared materials. Thirteen days after the second inoculation, all of the mice were challenged with 100 larvae, and 13 days later they were killed by cervical dislocation. The average number of worms recovered from the Group I mice was 38.8 (range: 28 to 48), the average for Group II was 55.4 (range: 48 to 62), and the average for Group III was 58.4 (range: 50 to 66). The average number of worms from Group I was significantly less than that from Group II (t value of 8.2; probability < 0.001). The averages for the two control groups were not significantly different.

As noted above, the crude aqueous extract of *N. dubius* larvae combined in equal parts with Freund's complete adjuvant caused a 29.9% reduction in worms compared with the adjuvant controls. Therefore, the effect of this preparation was more than twice as great as that reported for the one used by Van Zandt (1962, J. Parasit. **48:** 249–251). Moreover, only two injections rather than six were needed to produce this more effective response.

184

Demonstration of Immunity to *Nematospiroides dubius* in Recipient Mice Given Spleen Cells

Raymond Cypess

As first reported by Van Zandt (1961, J. Elisha Mitchell Sci. Soc. **77:** 300–309), mice given three spaced oral doses of 50 *N. dubius* larvae produce an immunity sufficient to cause a significant reduction of worms after a challenging infection. More recently, Lueker et al. (1968, J. Parasit. **54:** 1237–1238) showed that a single subcutaneous injection of 4,000 exsheathed larvae produces a strong immunity.

The results of studies on the mechanism of this immunity have led one worker to suggest that immediate hypersensitivity is involved, i.e., that reinfections initiate an anaphylactic reaction, and that changes associated with this prevent establishment of larvae (Panter, 1969, J. Parasit. **55:** 38–43). However, the failure to transfer the immunity with antiserum casts considerable doubt on the importance of humoral factors in causing the reduced worm burdens noted in immunized hosts. The fact that immunity has been transferred with lymphoid cells and not antiserum in studies of certain other helminths (Larsh, 1967, Am. J. Trop. Med. Hyg. **16:** 735–745) suggested that delayed rather than immediate hypersensitivity might be more important in the mechanism of immunity to *N. dubius*. Therefore, the following experiment was performed.

Ten male mice (12 weeks old) were selected as donors. These mice as well as the recipients are an isologous strain (C57). Each was injected with 4,000 exsheathed larvae intraperitoneally (IP) and 3 weeks later each was bled by cardiac puncture and then killed by cervical dislocation. The serum was collected in the conventional way, and the methods for collecting and calculating the number of spleen cells are described elsewhere (Larsh et al., 1969, J. Parasit. **55:** 726–729).

Forty male mice (12 weeks old) served as recipients and were separated into four equal groups. Those of Group I were injected IP with 0.5 ml of antiserum on the day after death of the donors, those of Group II were injected IP with 5×10^6 spleen cells at the same time, and those of Group III also were injected with the same number of spleen cells

after the cells were subjected to freezing and thawing according to the methods of Lawrence (1955, J. Clin. Invest. **34:** 219–230). The latter group was included as antigen controls, since this processing of the cells destroys them but does not prevent the passive transfer of antigens. The Group IV mice served as regular controls given no treatment prior to infection.

Ten days after the transfer of antiserum and cells, all 40 mice were challenged with 200 larvae. At 16 days after infection, the average number of worms recovered from the spleen cell recipients (49.6; range: 32 to 70) was significantly lower than that (95.4; range: 87 to 105) recovered from the regular controls (*t* value: 12.53; probability less than 0.001). On the other hand, the average numbers of worms recovered from the antiserum recipients (88.5; range: 69 to 115) and the frozen cell recipients (94.9; range: 87 to 109) were not significantly different from the number harbored by the regular controls. Likewise, there was not a significant difference between the numbers of worms harbored by the antiserum and frozen cell recipients.

It is clear from these results that the donor spleen cells did not transfer antigens to the recipients that resulted in the production of active immunity. However, in view of the fact that cells from donors sensitized to *Trichinella spiralis* antigens have been shown under certain conditions to produce both delayed hypersensitivity and, later, humoral antibodies (Kim et al., 1967, J. Immunol. **99:** 1156–1161), the results of the present experiment do not rule out the possibility that humoral antibody production during the 26 days after transfer of the cells played a role in the demonstrated significant reduction of worms in the spleen cell recipients.

In view of the above, and despite the failure of antiserum to affect the worm burden, further work is needed to rule out humoral antibody as an important contributing factor in this immunity. As the first step, it must be determined whether the recipients have detectable levels of circulating antibodies. In the meantime, with the information at hand, and in view of findings in studies with *T. spiralis* in mice (Larsh, 1967, Am. J. Trop. Med. Hyg. **16:** 123–132), the evidence favors the hypothesis that delayed hypersensitivity plays an important role in the immunity demonstrated against *N. dubius*.

JUVENILE NEMATODES (*ECHINOCEPHALUS PSEUDOUNCINATUS*) IN THE GONADS OF SEA URCHINS (*CENTROSTEPHANUS CORONATUS*) AND THEIR EFFECT ON HOST GAMETOGENESIS

J. S. PEARSE AND R. W. TIMM

Although sea urchins have been used as biological research material for many years, particularly as a source of gametes, there have been few reports of infection of these animals by nematodes. The nematode *Echinomermella grayi* (Gemmill and von Linstow, 1902) tentatively placed in the Mermithoidea by Chitwood (1933), is known from a single specimen found in the perivisceral coelom of *Echinus esculentus* off Britain, and perhaps a second specimen (Irving, 1910; Ritchie, 1910). Juvenile specimens of gnathostomatid nematodes have been reported twice from sea urchins: *Echinocephalus uncinatus* "perhaps accidentally" in tropical sea urchins (Shipley and Hornell, 1904) and a single specimen of *E. pseudouncinatus* in the gonads of *Arbacia punctulata* at Woods Hole, Massachusetts (Hopkins, 1935; Millemann, 1951). Nematodes have also been found in the gonads of the urchin *Astropyga pulvinata* off Acapulco, Mexico, but these have not been identified (J. S. Pearse, unpublished observations). Neither Hyman (1951, 1955), Johnson (1968), Johnson and Chapman (1969), nor Holland and Holland (1970) cite any other record of nematodes in echinoids.

We report herein the regular occurrence of juvenile *Echinocephalus pseudouncinatus* in the gonads of *Centrostephanus coronatus* off Southern California. Between about 40 and 80% of the sea urchins contained the nematodes in their gonads. Moreover, gametogenesis in parts of the gonads was profoundly affected by the presence of the nematodes.

MATERIALS AND METHODS

Most specimens of *Centrostephanus coronatus* were collected from 3 to 10 m depth by scuba diving off the northeastern shore of Big Fisherman's Cove, Santa Catalina Island, near the Santa Catalina Marine Biological Laboratory. Other samples were taken from Pin Rock in Catalina Harbor on the opposite side of Santa Catalina Island, and from Whistler's Reef off the mainland coast of California near Corona del Mar. These three areas are all in quite different waters: Big Fisherman's Cove is on the Santa Catalina Channel side of Santa Catalina Island; Catalina Harbor, although less than 3 km distant, is on the Pacific side of the island; and Whistler's Reef is about 25 km away on the opposite side of Santa Catalina Channel. Moreover, one sample of gonads from *C. coronatus* was taken on 26 November 1968 from Bahia Tortola (Turtle Bay) on the west-central coast of Baja California, Mexico.

The sea urchins were lodged deep in crevices among the rocks during the day and were extracted with the aid of a bent metal rod. During the night, animals foraged in the open and were more easily collected.

The specimens were dissected within a day after collection and all the gonads were carefully inspected for the presence of nematodes. Records were kept of the presence or absence of worms in the gonads, with notation of their relative abundance.

Gonads with nematodes were fixed in Bouin's solution or warm formalin-alcohol-acetic acid-water solution (FAA; 1:5:1:4). Some of the fixed specimens were dehydrated in an acetone series, cleared in benzene, embedded in paraffin, sectioned at 10 μ, and stained in hematoxylin and eosin. Others were dissected whole from the gonads.

TABLE I

Incidence of specimens of Centrostephanus *infected with* Echinocephalus

Date	Number of *Centrostephanus*	Number (and per cent) infected with *Echinocephalus*
Big Fisherman's Cove, Santa Catalina Island		
11 Aug. 69	38	26 (68.5)
20 Aug. 69	7	5 (71.5)
26 Aug. 69	18	14 (78.0)
31 Aug. 69	17	14 (82.5)
15 Oct. 69	10	9 (90.0)
29 Nov. 69	20	16 (80.0)
2 Feb. 70	18	13 (65.0)
9 Mar. 70	25	21 (80.0)
Catalina Harbor, Santa Catalina Island		
28 Aug. 69	22	9 (41.0)
Whistler's Reef, Corona del Mar		
10 Nov. 69	20	8 (40.0)
20 Nov. 69	18	7 (39.0)

OBSERVATIONS

Frequency of nematode infection in Centrostephanus coronatus

A total of 153 specimens of *C. coronatus* were collected from Big Fisherman's Cove, Santa Catalina Island, and examined for the presence of nematodes in their gonads. Of these, 118 (78%) were infected (Table I). The percentage infected in each sample ranged from 65 to 90%. All of the urchins in the August samples were sexed by microscopic examination of gonadal smears; there were 46 males and 34 females, and 31 and 28 were infected, respectively.

All the samples of *C. coronatus* from both Pin Rock in Catalina Harbor and Whistler's Reef off Corona del Mar had lower incidences of infection than those from Big Fisherman's Cove (Table I). Moreover, not only were a smaller percentage of animals infected, but those that were contained fewer nematodes. Most of the infected urchins in Big Fisherman's Cove had multiple infections, with several nematodes in each of the five gonads. In contrast, most of the infected urchins from both Pin Rock and Whistler's Reef had only one or two nematodes in only one or two of the gonads, and detection of the infection required careful dissection of all the gonads.

Seventeen specimens of *C. coronatus* were also collected and dissected on 26 November 1968 from Bahia Tortola (Turtle Bay) on the west-central coast of Baja California. At the time of the collection, the presence of nematodes was not known, but during the dissection, one gonad was found with an encysted nematode.

The juvenile nematode

The nematodes we found in the gonads of *C. coronatus* are all juveniles. They do not differ in any respect from Millemann's (1951) description of *Echinocephalus pseudouncinatus*. The most significant feature for identification is the structure of the head bulb (Fig. 1A), which bears 6 rows of about 40 hooks each, with a lateral separation of the hooks and smaller hooks on the dorsal and ventral areas. Figure 1B shows the tail, which was not drawn by Millemann (1951).

Echinocephalus pseudouncinatus was originally described from numerous juvenile specimens found in the foot of pink abalones *Haliotus corrugata* from San Clemente Island off Southern California (Millemann, 1951). This is the only species of *Echinocephalus* presently known from Southern California. Juveniles of another species of *Echinocephalus*, *E. uncinatus*, are known to infect molluscs elsewhere (Shipley and Hornell, 1904; Baylis and Lane, 1920), and it is likely that *E. pseudouncinatus* juveniles regularly infect both molluscs and *C. coronatus* in Southern California. Among the sea urchins of Southern California, however, juvenile *E. pseudouncinatus* seem to infect *C. coronatus* specifically. Hundreds of specimens of the sea urchins *Strongylocentrotus purpuratus*, *S. franciscanus*, and *Lytechinus anamesus* from Southern California have been dissected during the past year and only a single specimen of *E. pseudouncinatus* has been found in the gonad of one *S. purpuratus* collected near San Diego. The single, remarkable record of what seems to be a juvenile specimen of *E. pseudouncinatus* in the gonad of a specimen of *Arbacia punctulata* at Woods Hole, Massachusetts, should again be noted (Hopkins, 1935; Millemann, 1951).

Possible adult hosts

Adults of the genus *Echinocephalus* inhabit the intestine, usually the spiral valve, of elasmobranchs (Hyman, 1951; Yamaguti, 1961). The most conspicuous elasomobranch in the vicinity of Big Fisherman's Cove, Santa Catalina Island, is the California horned shark *Heterodontus francisci*. This species is known to feed on sea urchins and its teeth and bones often are colored purple, presumably from an accumulation of sea urchin naphthoquinone pigments (Leighton Taylor, Scripps Institution of Oceanography, personal communication). A similar purple coloration of teeth and bones occurs in the sea otter *Enhydra lutris*, which feeds on sea urchins (Fox, 1953). Other species of horned sharks are known to feed on sea urchins, including *H. phillipi* and *H. galeatus* off southeast Australia, and *H. japonicus* off Japan (Smith, 1942). Moreover, Saville-Kent (1897, page 192) noted the teeth of *H. phillipi* ". . . are not infrequently stained a deep purple, through constant indulgence in a dietary of the commoner purple urchin."

189

Accordingly, three specimens of California horned sharks were collected in February 1970 from Big Fisherman's Cove, Santa Catalina Island, and the intestinal contents examined. One large female, measuring 76 cm total length,

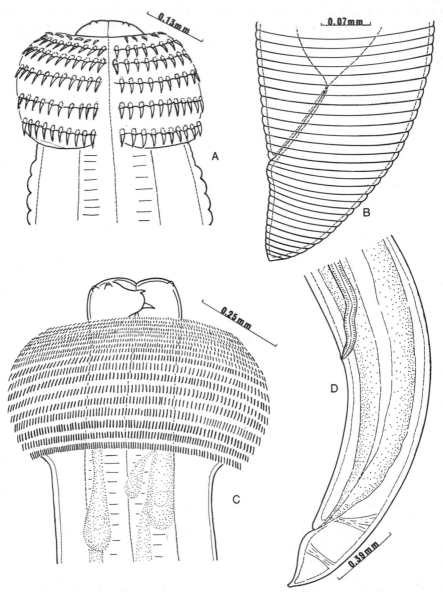

FIGURE 1. Juvenile head, lateral view (A) and juvenile tail (B) of *Echinocephalus pseudouncinatus* from the gonad of *Centrostephanus coronatus,* and female head, lateral view (C) and female tail (D) of *Echinocephalus* sp. from the spiral valve of *Heterodontus francisci.*

contained three adult specimens of *Echinocephalus* in its spiral valve. The adult worms, however, have 18 rows of hooks on the head bulbs, each containing 150–200 hooks and therefore, by current taxonomic criteria, cannot be considered the same species as the juveniles in the sea urchin. The number and arrangement of hooks on the head bulb is the same as in *E. southwelli* Baylis and Lane, 1920, but the vulva and cervical sacs are much more anterior than in that species. Since we found no males, no attempt is made to describe the nematode as a new species, but the measurements, a brief description, and drawings (Fig. 1C, D) are given in order that parasitologists might be aware of this previously unknown host-parasite relationship.

Young entire female: Length = 22.46 mm; maximum body diameter = 0.68 mm; esophagus = 3.48 mm; esophagus-vulva = 17.42 mm; vulva-anus = 1.19 mm; tail = 0.37 mm; anal body diameter = 0.26 mm; head bulb = 0.75 × 0.44 mm; lip region = 0.26 × 0.12 mm.

Anterior of one female: Length = 17.42 mm; maximum body diameter = 1.2 mm; head bulb = 0.93 × 0.17 mm; lip region = 0.44 × 0.26 mm.

Posterior of two females: Length = 19.86–24.0 mm; maximum body diameter = 1.34 mm; vulva-anus = 2.75–3.67 mm; tail = 0.58–0.70 mm; anal body diameter = 0.16 mm.

Description: Cuticle sometimes coarsely but irregularly annulated, probably due to contraction, each annulus bearing finer striations. Eighteen rows of hooks on the head bulb, composed of 150–200 hooks each, larger hooks toward posterior; distance between rows less than length of hooks; maximum length of hooks 0.035 mm. Two large lips, each bearing two prominent papillae; overlapping lobe of lip bears two toothlike projections. Cervical sacs extending about ½ length of esophagus in young female. Deirids not observed. Tail broadly conical, narrowly rounded at tip. These adult nematodes, as well as a representative collection of juvenile *E. pseudouncinatus* from *C. coronatus,* are held in the Nematode Collection of the University of California, Davis.

Nematode effect on urchin gonads

The echinoid gonad consists of five primary tubules that grow orally along the interambulacrals from each of the aborally situated gonopores (Fig. 2). The five primary tubules branch repeatedly so that in adults the five gonads are large organs of intertwining tubules. In *C. coronatus* the gonads are anchor-shaped, with the "flukes" of the anchor situated orally. The gonads are enveloped completely with the thin, ciliated epithelium of the perivisceral coelom, and strands of this epithelium secure the gonad to the inner surface of the test wall and to the gut. Under the perivisceral coelomic epithelium, thin layers of smooth muscle and connective tissue cover the inner germinal epithelium. The germinal epithelium is derived from the hemal strand and consists of both reproductive cells (gonials, spermatocytes or oocytes, spermatozoa or ova) and accessory cells (nutritive phagocytes). Often there is a space ("perihemal coelom") between the perivisceral coelomic epithelium and the muscle layer and another space ("hemal sinus") between the muscle layer and the germinal epithelium.

Young juvenile nematodes were lodged between the perivisceral coelomic epithelium and the germinal epithelium; that is, within the hemal or perihemal

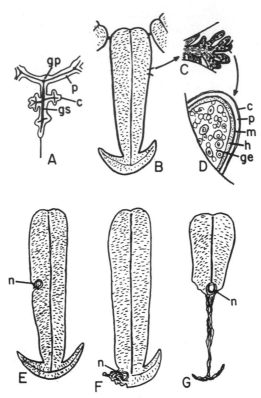

FIGURE 2. Diagrams showing the structure of sea urchin gonads during early development in a juvenile urchin (A) (after Tahara and Okada, 1968), when fully developed in an adult *Centrostephanus coronatus* (B, C, D), and the effect on the gonad when infected by juvenile *Echinocephalus pseudouncinatus* encysted near the end of a peripheral tubule (E), in the main tubule leading to one of the flukes of the gonad (F), and in the main tubule (gonoduct) near the middle of the gonad (G). All of the gonads are shown from their perivisceral coelomic sides with the oral portion down; abbrev.: c, perivisceral coelomic epithelium; ge, germinal epithelium; gp, position of gonopore; gs, hemal strand; h, "hemal sinus"; m, muscle; n, encysted nematode; p, "perihemal coelom."

spaces of the gonadal wall. Larger worms displaced most of the germinal epithelium in the infected tubule and thick, fibrous connective tissue encysted both the worm and associated germinal epithelium. Cysts with large juveniles protruded into the perivisceral coelom; large juveniles, probably nearly full-grown, broke through the wall of the gonadal cyst and extended free into the coelom.

Although gonadal tubules adjacent to those with the worms usually contained well-developed spermatogenic or oogenic cells, most of the tissue within the cysts in the infected tubules consisted of nongametogenic accessory or connective cells. Moreover, gametogenesis within infected tubules distal (oral) to the worm also seemed suppressed. When the infection occurred near the end of a side tubule, as was usually the case, there was little effect on the overall gametogenic state of the gonad (Fig. 2E). Occasionally, however, infection occurred within a main tubule and, in such cases, the gonads were shriveled distal to the sites of infection (Fig. 2F, G). In one dramatic example, two worms had lodged in the main tubule (probably the gonoduct) in the middle of one testis; spermatogenesis appeared

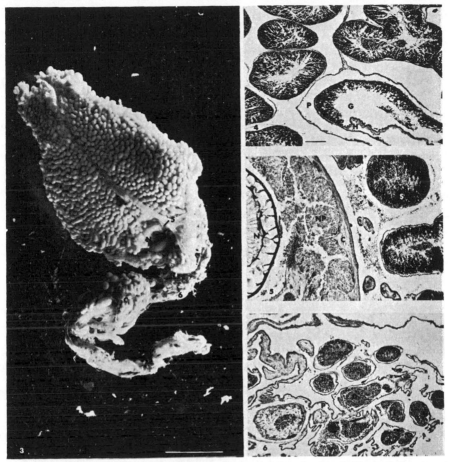

FIGURE 3. A testis of *Centrostephanus coronatus* infected with *Echinocephalus, pseudo-uncinatus* showing a cyst containing one nematode (C) and a second nematode broken partially free of its cyst (N). The oral portion of the gonad is at the bottom of the photograph; note the normal-appearing aboral half of the gonad and the shriveled oral portion. Numbers 4, 5, and 6 correspond to Figures 4 through 6; the scale line indicates 5 mm.

FIGURE 4. Section through the normal-appearing aboral portion of the testis in Figure 3, showing the gonoduct (G) surrounded by perithemal (P) and hemal (H) spaces, and the testicular tubules full of spermatogenic cells (S).

FIGURE 5. Section through the nematode in the testis in Figure 3, showing a portion of the nematode (N) and extensive nutritive phagocytic tissue (NP) in a tubule surrounded by a fibrous cyst wall (C), and adjacent tubules full of spermatogenic cells (S).

FIGURE 6. Section through the shriveled oral portion of the testis in Figure 3 just below the nematode, showing the germinal epithelium full of degenerating cells (D) and large perihemal spaces (P). Figures 4–6 are from 10 μ paraffin sections stained with hematoxylin and eosin; all are at the same magnification and the scale line in Figure 4 indicates 100 μ.

active and normal in the aboral half of the testis, while the tubules were shrunken and filled with degradation products orally (Figs. 3–6).

The finding that encysted juvenile worms can suppress urchin gametogenesis has important implications regarding gametogenic control. Suppression does not seem to be due to the release and diffusion of some substance by the worms because gametogenesis is suppressed only oral to the infections rather than all around them. Moreover, the worms are coiled and encased in cysts produced by the urchin gonads; they are not likely to move through the gonads and disrupt gametogenesis. Rather, the encysted worms seem to block the passage of some material that is essential for gametogenesis. Such a gametogenic regulating substance, although never directly demonstrated, is almost certainly present because gametogenesis occurs in synchrony among all five gonads in individual sea urchins (Pearse, 1969). The encysted nematodes probably do not block transport of simple nutrients; radioactive tracing studies have shown that nutrient transfer in urchins occurs mainly through the perivisceral coelom (Farmanfarmaian and Phillips, 1962). Instead, the apparent blockage of gametogenesis by the juvenile nematodes indicates the presence of a hormonal substance which regulates gametogenesis in sea urchins and is transported *within* the gonad (perhaps through the perihemal or hemal spaces) rather than through the perivisceral coelomic fluid.

We are indebted to Dr. Russel L. Zimmer, Resident Director, for providing aid and facilities for these studies at the Santa Catalina Marine Biological Laboratory, Santa Catalina Island, and to Dr. Nicholas D. Holland, Dr. W. Duane Hope, Dr. Phyllis T. Johnson, and Mr. Leighton Taylor for advice and criticism. This study was supported in part by the Federal Water Quality Administration Grant No. 18050 DNV. The sample from Bahia Tortola, Baja California, was taken during Stanford Oceanographic Expedition, TE VEGA Cruise 20, supported by the National Science Foundation Grant Nos. GB6870 and GB6871. Part of the manuscript was prepared while the senior author was a visting faculty member of the University of California, Santa Cruz.

Summary and Conclusions

1. The juvenile phase of the nematode *Echinocephalus pseudouncinatus* occurs commonly in the gonads of the sea urchin *Centrostephanus coronatus* off Southern California. The gonads of about 78% of the urchins in Big Fisherman's Cove, Santa Catalina Island, were heavily infected with the juveniles. About 40% of the urchins at Pin Rock in Catalina Harbor, Santa Catalina Island, and at Whistler's Reef off Corona del Mar on the mainland coast of California were infected, although the infection was usually not as severe as in Big Fisherman's Cove. A juvenile, probably of *E. psuedouncinatus,* also was found in gonads of *C. coronatus* collected from Bahia Tortola off west-central Baja California. Although juveniles of *E. pseudouncinatus* also occur commonly in the foot of pink abalones, they only rarely infect other species of sea urchins in Southern California.

2. The California horned shark *Heterodontus francisci* seems a likely host of the adult phase of *E. pseudouncinatus*. However, when specimens of the horned

shark were examined, adult specimens of *Echinocephalus* were found which do not seem to be the same species as our juvenile specimens.

3. The juvenile nematodes were encysted mainly in the spaces (perihemal or hemal) between the perivisceral coelomic epithelium and the germinal epithelium of the *Centrostephanus* gonad. Large juveniles filled the host gonadal tubule and bulged into the perivisceral coelom. Host gametogenesis was suppressed in the infected gonadal tubule, especially in the oral parts of such tubules. Gametogenesis in adjacent tubules did not seem affected. When the juveniles were in major tubules of the host gonad, severe suppression of host gametogenesis occurred in the oral parts of the gonad. It is suggested that encysted juveniles block the passage through the gonadal tubules of some hormonal substance that regulates urchin gametogenesis.

LITERATURE CITED

BAYLIS, H. A., AND C. LANE, 1920. A revision of the nematode family Gnathostomidae. *Proc. Zool. Soc. London,* **1920**: 245–310.

CHITWOOD, B. G., 1933. The systematic position of *Echinonema grayi* Gemmill, 1901. *J. Parasitol.,* **20**: 104.

FARMANFARMAIAN, A., AND J. H. PHILLIPS, 1962. Digestion, storage, and translocation of nutrients in the purple sea urchin (*Strongylocentrotus purpuratus*). *Biol. Bull.,* **123**: 105–120.

FOX, D. L., 1953. *Animal Biochromes.* Cambridge University Press, Cambridge, 379 pp.

GEMMILL, J. F., AND O. VON LINSTOW, 1902. *Icthyonema grayi. Arch. Naturgesch.,* **68**: 113–118.

HOLLAND, N. D., AND L. Z. HOLLAND, 1970. A bibliography of echinoderm biology, continuing Hyman's 1955 bibliography through 1965. *Pubbl. Staz. Zool. Napoli,* **37**: in press.

HOPKINS, S. H., 1935. A larval *Echinocephalus* in a sea urchin. *J. Parasitol.,* **21**: 314–315.

HYMAN, L. H., 1951. *The Invertebrates. Volume III. Acanthocephela, Aschelminthes and Entroprocta.* McGraw-Hill Book Co., New York, 572 pp.

HYMAN, L. H., 1955. *The Invertebrates. Volume IV. Echinodermata.* McGraw-Hill Book Co., New York, 763 pp.

IRVING, J., 1910. Nemertine within test of sea-urchin. *The Naturalist (London),* **1910**: 6.

JOHNSON, P. T., 1968. *An Annotated Bibliography of Pathology in Invertebrates other than Insects.* Burgess Publishing Co., Minneapolis, Minnesota, 322 pp.

JOHNSON, P. T., AND F. A. CHAPMAN, 1969. *An annotated bibliography of pathology in invertebrates other than insects supplement.* Misc. Publ. No. 1, Center for Pathobiology, University of California, Irvine, 76 pp.

MILLEMANN, R. E., 1951. *Echinocephalus pseudouncinatus* n. sp., a nematode parasite of the abalone. *J. Parasitol.,* **37**: 435–439.

PEARSE, J. S., 1969. Reproductive periodicities of Indo-Pacific invertebrates in the Gulf of Suez. I. The echinoids *Prionocidaris baculosa* (Lamarck) and *Lovenia elongata* (Gray). *Bull. Mar. Sci.,* **19**: 323–350.

RITCHIE, J., 1910. Worm parasitic in sea-urchin. *The Naturalist (London),* **1910**: 94.

SAVILLE-KENT, W., 1897. *The Naturalist in Australia.* Chapman and Hall, London, 302 pp.

SHIPLEY, A. E., AND J. HORNELL, 1904. The parasites of the pearl oyster. *Rept. Govt. Ceylon Pearl Oyster Fish. Gulf Manaar* (Herdmann, London), **2**: 77–106.

SMITH, B. G., 1942. The heterodontid sharks, their natural history and external development of *Heterodontus (Cestracion) japonicus* based on notes and drawings by Basford Dean. Pages 647–770 in E. W. Gudger, Ed., *The Basford Dean Memorial Volume, Archaic Fishes, Part II, Article VIII,* American Museum of Natural History, New York.

TAHARA, Y., AND M. OKADA, 1968. Normal development of secondary sexual characters in the sea urchin, *Echinometra mathaei. Publ. Seto Mar. Biol. Lab.,* **16**: 41–50.

YAMAGUTI, S., 1961. *Systema Helminthum. Volume III. The Nematodes of Vertebrates. Parts 1 and 2.* Interscience Publishers, New York, 1261 pp.

I regret that we missed the following important paper: Raymond E.
Millemann, 1963. Studies on the taxonomy and life history of echino-
cephalid worms (Nematoda:Spiruroidea) with a complete description
of Echinocephalus pseudouncinatus Millemann, 1951. J. Parasitol.,
49:754-764. Millemann described in detail third-stage, molting
third-stage and fourth-stage juveniles, and young and mature adults
of Echinocephalus pseudouncinatus collected from the spiral valves
of the horned shark Heterodontus francisci and the bat stingray
Myliobatis californicus from the Gulf of California. During the
third molt the number of rows of hooks on the head of E. pseudouncinatus increases from 8 to approximately 24, and the number of
hooks per row increases from about 40 to about 300. When the
fourth-stage juvenile molts to the young adult, there is a slight
reduction in the number of hooks, and in the adult there are about
19 rows of hooks, each row with about 150-200 hooks.

From Millemann's paper we can now say that the adult specimens of
Echinocephalus that we collected from the spiral valve of Heterodontus francisci from Big Fisherman's Cove, Santa Catalina Island,
were, in fact, E. pseudouncinatus. Our suspicion that H. francisci
might be the adult host of E. pseudouncinatus is thus confirmed.
The sharks are probably infected by eating the encysted juveniles
in the gonads of Centrostephanus coronatus. Millemann suggests
that only two hosts are necessary for E. pseudouncinatus: elasmobranchs (horned sharks, stingrays) as the definitive adult host,
and molluscs (pink abalone) as the obligatory intermediate juvenile
host. From our studies, it appears that the sea urchin C. coronatus
also can be an important intermediate host. In this respect, it is
of interest that pink abalone do not occur in the Gulf of California
while C. coronatus is common there.

John S. Pearse

IN VITRO DETECTION OF HOMOCYTOTROPIC ANTIBODY IN LUNGWORM-INFECTED ROCKY MOUNTAIN BIGHORN SHEEP

R. J. HUDSON, P. J. BANDY AND W. D. KITTS

INTRODUCTION

Immediate-type hypersensitivity is commonly associated with parasitic infections (Ogilvie, 1964; Hogarth-Scott, Johansson & Bennich, 1969; Barratt & Herbert, 1970; McAninch & Patterson, 1970). Antibodies mediating this reaction have been termed homocytotropic since they sensitize only homologous tissues for the anaphylactic release of histamine and other vasoactive compounds. In man, sensitization has been attributed to a distinct

197

immunoglobulin class called IgE (Ishizaka & Ishizaka, 1967, 1968; Johansson & Bennich, 1967). Similar antibodies have been detected in animals; however, their analogy to the homocytotrophic antibody of man has been established only in certain species (Strejan & Campbell, 1968; Freeman *et al.*, 1969; Patterson, 1969).

Peripheral leucocytes adsorb homocytotropic antibody from allergic serum (Augustin, 1963). The non-specific uptake of other immunoglobulins by these cells is lower than lung or dermal tissues (Parish, 1970). Several *in vitro* assays based on leucocyte sensitization have been developed (Levy & Osler, 1967; Fitzpatrick *et al.*, 1967).

Homocytotropic antibodies, which sensitize polymorphonuclear leucocytes for adherence to the parasite cuticle, were detected in lungworm infections (*Protostrongylus stilesi*) of the Rocky Mountain bighorn, a native North American wild sheep. By measuring the adherence of passively sensitized cells to larvae in the presence and in the absence of serum from infected animals, both blocking and homocytotropic activity could be determined. This study was conducted to evaluate this reaction as an *in vitro* correlate of immediate-type hypersensitivity in *Protostrongylus* infections.

MATERIALS AND METHODS

Animals

Four Rocky Mountain bighorn sheep (*Ovis canadensis canadensis* Shaw), captured in the East Kootenay region of British Columbia, were maintained at the University wildlife unit. These animals harboured light natural infections of several parasitic helminths but the most predominant was the parenchymal lungworm (*Protostrongylus stilesi* Dikmans). Domestic sheep (*O. aries*) were used as recipients in the skin test experiments as well as cell donors for *in vitro* sensitization.

Antigens

Lungworm lesions were excised from the lungs of naturally infected wild bighorns. The tissue was teased apart in saline and filtered through cheesecloth to remove the larger lung fragments. A purified suspension containing approximately 20% ova and 80% larvae was separated from lighter tissue debris by repeated sedimentation in normal saline.

Passive transfer of cutaneous hypersensitivity (P-K reaction)

Skin sensitization was conducted according to Hogarth-Scott (1969). Parasite allergen (0·1 ml representing approximately 800 larvae) was injected into the prepared sites after the intrajugular infusion of 5 ml of 1·5% trypan blue dye. The diameter of extravasated dye was recorded after 30 min.

Preparation of leucocyte suspensions

Two volumes of heparinized blood were added to one volume of 4% gelatin (Difco) dissolved in saline buffered to pH 7·2 with 0·1 M sodium phosphate (PBS). After gentle mixing, the tubes were incubated in a vertical position at 37°C for 40 min to permit sedimentation of erythrocytes. The leucocyte-rich plasma was collected and the cells were washed with 20 ml PBS containing 1·5% polyglucose (Ficoll, Pharmacia, Uppsala). Hypotonic lysis was employed for the purification of cell suspensions from domestic sheep since their erythrocytes would not sediment in gelatin or dextran.

In vitro sensitization

Washed peripheral leucocytes were suspended in 1 ml of allergic serum and incubated at 37°C for 3 hr. Following sensitization, the cells were washed twice with 20 ml of PBS–Ficoll.

Washed-cell test

To determine homocytotropic activity, washed sensitized cells (6×10^6/ml) were examined for adherence to larvae in a medium containing normal rabbit serum diluted with an equal volume of PBS. Parasite allergen (approximately 450 larvae) was added to 0·8 ml of the sensitized cell suspension. Following incubation (1 hr at 22°C) the sediment was examined microscopically and the leucocytes adhering to the ova and larvae were enumerated.

Decomplemented-serum test

Effective hypersensitivity, accounting for the influence of both homocytotropic and blocking antibodies, was assayed by measuring the adherence of normal or desensitized cells in a 0·3 ml suspension when incubated with larvae in 0·4 ml of allergic test serum. Blocking activity could then be estimated by the relative difference between the washed-cell and the decomplemented-serum test. Since neutrophils possess receptors for fixed complement (see Cochrane, 1968), test sera were decomplemented by absorption with zymosan (5 mg/ml) prior to the adherence test (Henson, 1969).

Elution of homocytotropic antibody from sensitized leucocytes

Sensitized leucocytes were suspended in PBS containing 0·004 M EDTA and incubated for 6–8 hr at 37°C with occasional changes of the suspending buffer. The supernatants were pooled, concentrated by ultra-dialysis (Diaflo, Amicon Corp.), and tested by micro-immunoelectrophoresis (Scheidegger, 1955).

Effect of immunoglobulin-class specific antisera on cell adherence

The preparation of antisera to the ovine immunoglobulins, IgA, IgM, $7S\gamma_1$ and $7S\gamma_2$, has been described previously (Hudson, Bandy & Kitts, 1970). To prevent sensitization by the immunoglobulin fractions used for absorption, IgG from rabbit antiserum was prepared by chromatography on DEA-Sephadex A-50 (Pharmacia, Uppsala) using phosphate buffer (0·02 M, pH 6·5). Sensitized cells were incubated in these antisera for 30 min at 37°C, then washed with 20 ml PBS and subjected to the adherence test.

Estimation of orosomucoid

The levels of circulating orosomucoid were estimated by disc electrophoresis of the perchloride acid-soluble serum proteins. The results were expressed on a relative basis as the seromucoid index (Hudson, Kitts & Bandy, 1970a).

RESULTS

Neutrophils and eosinophils, passively sensitized with allergic serum or obtained from parasitized animals, adhered to parasite larvae in a medium with strong dielectric properties provided by normal serum or dextran (Fig. 1). Heterologous granulocytes (rabbit) were not sensitized for adherence. The sensitivity of the washed-cell test was directly related

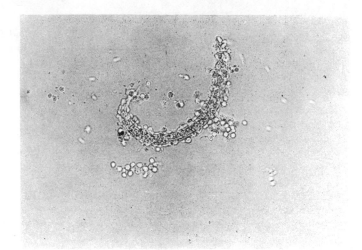

Fig. 1. Peripheral leucocytes sensitized with ovine homocytotropic antibody adhering to *Protostrongylus* larvae *in vitro*.

to the dielectric constant. Highly sensitized cells taken from bighorns during periods of increased hypersensitivity adhered to larvae *in vitro* even in the absence of strongly polar molecules.

The use of mixed suspensions of ova and larvae, obtained directly from the lung lesion, permitted an evaluation of the specificity of the hypersensitive response. Adherence to ova

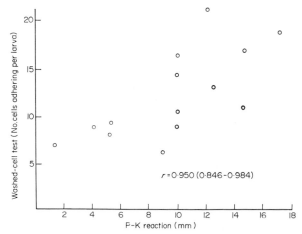

Fig. 2. Correlation between the washed-cell test measured as the number of passively sensitized polymorphs adhering per larva, and the diameter of the skin reaction (P–K reaction) elicited 72 hr after passive transfer of allergic serum. Serum samples for this comparison were collected from three animals at a number of representative intervals throughout the seasonal parasite cycle.

and larvae was independent. In addition, two antigenic populations of larvae were present in one of the antigen suspensions used in this study. Sensitized cells from captive animals consistently adhered to only 40% of the larvae present even when the adherence reaction to this fraction was high. Rarely, serum samples from wild bighorns would sensitize cells for adherence to all larvae present in the suspension.

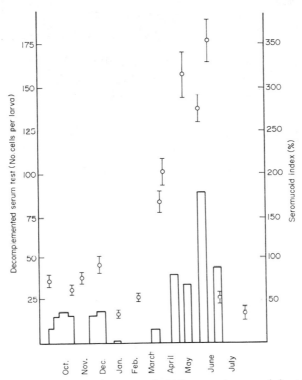

FIG. 3. The effective hypersensitive response of a bighorn ewe to seasonal changes in parasite activity. The points represent the mean number of cells adhering per larva in the decomplemented-serum test (accounting for both homocytotropic and blocking antibodies). The vertical lines represent the standard error. Adherence was evaluated for approximately thirty larvae in each case. The bar graph shows the relative level of orosomucoid, a measure of inflammation, in serum from the infected animal.

Correlation between in vitro cell adherence and hypersensitivity

Homocytotropic activity, evaluated by the washed-cell test, was directly correlated with 72-hr skin sensitizing activity (Fig. 2).

Since the animals used in this study carried very light natural infections, the *in vitro* assays of immediate hypersensitivity could not be related to gross clinical symptoms. Orosomucoid, a sensitive index of inflammation (Dearing, McGuckin & Elveback, 1969; Werner, 1969), exhibits a marked response to seasonal changes in parasite activity (Hudson *et al.*, 1970a). The decomplemented-serum test, which measures the net result of both

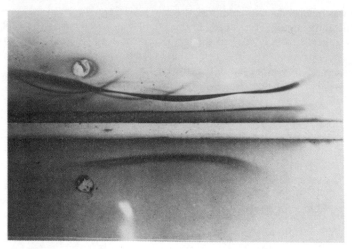

FIG. 4. Immunoelectrophoresis of a $7S\gamma_1$ immunoglobulin eluted from sensitized peripheral leucocytes (lower well). Whole serum as a control was placed in the upper well. The anode is located at the left side of the figure.

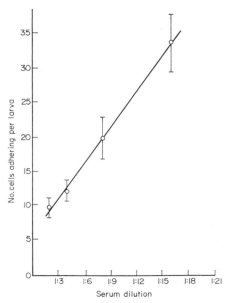

FIG. 5. Effect of cell adherence of dilution of allergic serum prior to the decomplemented-serum test. Each point shows the mean and standard error for cell adherence to approximately thirty larvae.

homocytotropic and blocking activity, was closely associated with these inflammatory changes (Fig. 3).

Relation of cell sensitization to the immunoglobulin classes

The immunoglobulin mediating the adherence of washed peripheral leucocytes to larvae appeared to be $7S\gamma_1$. The concentrated supernates from cells desensitized by incubation in PBS–EDTA, revealed the presence of this immunoglobulin when tested by immuno-electrophoresis (Fig. 4). In addition, antisera to $7S\gamma_1$, but not to IgA, IgM and $7S\gamma_2$, caused a reduction in cell adherence. Absorption of the inhibiting antiserum with colostrum, which contains high levels of $7S\gamma_1$ (Heimer, Jones & Maurer, 1969), abolished its activity.

A $7S\gamma_1$ fraction obtained from allergic serum by rivanol fractionation and subsequent anion exchange chromatography, retained the ability to sensitize ovine skin for anaphylaxis The ion exchange step was accompanied by a partial loss of homocytotropic activity.

Blocking activity

Both $7S\gamma_1$ and $7S\gamma_2$, purified from allergic serum by anion exchange chromatography, exhibited blocking activity when incubated with larvae prior to the washed-cell test. Dilution of allergic serum used in the decomplemented-serum test with normal serum produced marked increases in cell adherence (Fig. 5).

DISCUSSION

Immediate-type skin reactions are difficult to study in the sheep (Kaplan & Freeman, 1969). The adherence of sensitized polymorphonuclear leucocytes to ova and larvae provided a convenient alternative assessment of immediate hypersensitivity in the parasitic diseases of this species. This procedure was rapid and obviated many of the technical difficulties associated with other *in vitro* methods based on histamine release. However, like the 'double-layer leucocyte agglutination' test of Fitzpatrick *et al.* (1967), possible interference by non-anaphylactic cytophilic antibodies cannot be rigidly excluded. Apparently, such antibodies are not important in this parasitic infection since, in spite of difficulties quanti-tating cutaneous anaphylaxis, a satisfactory correlation was obtained between *in vitro* adherence and skin sensitizing activity.

Granulocyte adherence appeared to be specific and dependent on cytophilic antibody. The larval cuticle represents a rather poor substrate for non-specific granulocyte adherence. Jeska (1969) did not observe the adhesion of peritoneal polymorphonuclear leucocytes to parasitic worms. Soulsby (1967) and Morseth & Soulsby (1969) noted that adherence to *Ascaris suum* larvae was strictly dependent on the fixation of complement.

Coulombic forces were apparently important in adherence. The addition of molecules with large dipole moments tended to disperse the electrostatic repulsive barrier due to the negative charges on both polymorphs and larvae. Treatment of larvae with neuraminidase, which cleaves the highly anionic sialic acid residues from the cuticle, engendered marked increases in adherence. This effect was partially reversed by performing the reaction in the presence of proteins or dextran (unpublished results). When the degree of cell sensitization was high, adherence occurred even in simple electrolyte solutions. Possibly, the attachment of antibody tends to partially cancel the surface charge and thereby reduce repulsion.

Homocytotropic antibody is not strictly species specific in its activity since it retains the capacity to sensitize the tissues of closely related species. Sensitization of monkey tissues (Layton, 1965; Ishizaka *et al.*, 1970) has been used to study human reaginic antibodies. Both *in vitro* and *in vivo* passive sensitization were freely achieved between bighorn and domestic sheep in this study.

Adherence of peripheral granulocytes to parasitic larvae appeared to be mediated principally by a $7S\gamma_1$ immunoglobulin, although the presence of additional homocytotropic antibodies cannot be excluded. Pan *et al.* (1968) and Kaplan & Freeman (1969) reported that skin sensitizing activity in domestic sheep did not parallel the binding of radio-labelled antigen by any of the four major ovine immunoglobulins. However, Hogarth-Scott (1969) was able to block cutaneous anaphylaxis with IgG_1a, a newly reported sub-class of $7S\gamma_1$ occurring in the serum of some sheep. Antiserum to this subclass also elicited reversed passive cutaneous anaphylaxis.

Analogies among the homocytotropic antibodies of different species have been suggested on the basis of their heat sensitivity and persistence at the skin test site. In this respect, ovine homocytotropic antibody appeared similar to human IgE (Hogarth-Scott, 1969). In most other species, however, $7S\gamma_1$ antibodies are heat stable and sensitize tissues for short (hours) instead of long intervals (days) (Mota *et al.*, 1969; Revoltella & Ovary, 1969). Guinea-pig anaphylactic antibody demonstrates a comparable combination of characteristics (Bloch, 1967). However, a number of skin sensitizing antibody populations have been detected in experimental animals (Morse, Austen & Bloch, 1969; Oliveira *et al.*, 1970) and further studies may reveal a similar heterogeneity in the sheep.

Reversal of sensitization *in vitro* depends upon the concentration of competitive non-antibody immunoglobulin (Mongar & Winne, 1966; Binaghi, 1968). The persistence of passively transferred antibody may well be a function not only of the characteristics of the sensitizing immunoglobulin but also the concentration and turnover of non-antibody immunoglobulin in the skin.

Blocking activity could be easily demonstrated by diluting serum from parasitized animals prior to the decomplemented serum test. Dilution apparently does not affect both cytophilic and blocking antibody equally. Sufficient cell-sensitizing antibody may have been present to approach saturation of all cell receptors. This blocking activity was complex. Both $7S\gamma_2$ and $7S\gamma_1$ fractions from infected animals were capable of reducing the adherence of sensitized cells. Possibly, $7S\gamma_1$ consists of immunoglobulin populations which exhibit varying affinities for cells. Strongly cytophilic antibodies would function as homocytotropic antibodies. Populations which were more weakly cytophilic would exhibit blocking activity. Alternatively, more than one immunoglobulin may be involved. Blocking antibodies presumably compete with cell sensitizing antibodies for antigenic sites. Conversely, certain serum glycoproteins and perhaps antibodies may actually enhance adherence by masking highly anionic groups on the cuticle thereby reducing electrostatic repulsion. These aspects are presently being studied.

An explanation for the presence of two antigenic populations of *Protostrongylus* larvae is not certain. They may represent two larval stages since larvae hatched in the lung and larvae acquired by reinfection are difficult to distinguish morphologically. This would explain why cells from captive bighorns, which were free of reinfection, adhered to only one population whereas occasionally serum from wild bighorns, which are naturally exposed to infective larvae, would sensitize cells for adherence to both larval populations. Another explanation is that certain components of the cuticle may have been damaged during purification and storage. The difference between the responses would then be due to the recognition of different antigenic determinants.

Preliminary studies using cell adherence techniques in *Protostrongylus* infections indicate that low blocking activity may be as important as elevated homocytotropic antibody during the intense inflammatory reaction associated with the seasonal increases in parasite activity (Hudson, Kitts & Bandy, 1970b). Immunogenic inflammation certainly accompanies parasitism but it is difficult to evaluate its role in the parasite–host interaction. Although anaphylaxis may affect the parasite directly, Barth, Jarrett & Urquhart (1966) emphasized the importance of extravasated non-anaphylactic antibodies from serum. The ultimate expression of immunity will probably be found to involve a number of immunologic effector mechanisms directed against a number of parasite antigenic targets.

ACKNOWLEDGMENTS

This study was supported, in part, by the National Research Council of Canada, Grant A-132. We thank Miss V. I. Curylo for technical assistance.

REFERENCES

AUGUSTIN, R. (1963) Allergens and other antigens of grass pollens: their isolation, characterization and standardization in relation to immune responses evoked in animals and man. *Fifth European Congress of Allergy, Basel,* 1962, p. 137. Selbstverlag Shwerz. Allergieges Druck: Schwabe, Basel.

BARRATT, M.E.J. & HERBERT, I.V. (1970) Skin sensitizing antibody in experimental infections with *Metastrongylus* spp. (Nematoda). *Immunology,* 18, 23.

BARTH, E.E.E., JARRETT, W.F.H. & URQUHART, G.M. (1966) Studies on the mechanism of the self-cure reaction in rats infected with *Nippostrongylus brasiliensis. Immunology,* 10, 459.

BINAGHI, R.A. (1968) The sensitization of tissues, and interference of non-specific gamma-globulin. *Biochemistry of the Acute Allergic Reactions* (Ed. by K. F. Austen and E. L. Becker), p. 53. Blackwell Scientific Publications, Oxford and Edinburgh.

BLOCH, K.J. (1967) The anaphylactic antibodies of mammals including man, *Progr. Allergy,* 10, 84.

COCHRANE, C.G. (1968) Immunologic tissue injury mediated by neutrophilic leukocytes. *Advanc. Immunol.* 9, 97.

DEARING, W.H., McGUCKIN, W.F. & ELVEBACK, L.R. (1969) Serum α_1-acid glycoprotein in chronic ulcerative colitis. *Gastroenterology,* 56, 295

FITZPATRICK, M.E., CONNOLLY, R.C., LEA, D.J., O'SULLIVAN, S.A., AUGUSTIN, R. & MACCAULAY, M.B. (1967) *In vitro* detection of human reagins by double-layer leucocyte agglutination: methods and controlled blind study. *Immunology,* 12, 1.

FREEMAN, M.J., BRALEY, H.C., KAPLAN, R.M. & McARTHUR, W.P. (1969) Occurrence and properties of rabbit homocytotropic antibody. *Int. Arch. Allergy,* 36, 530.

HEIMER, R., JONES, D.W. & MAURER, P.H. (1969) The immunoglobulins of sheep colostrum. *Biochemistry,* 8, 3937.

HENSON, P.M. (1969) The adherence of leucocytes and platelets induced by fixed IgG antibody or complement. *Immunology,* 16, 107.

HOGARTH-SCOTT, R.S. (1969) Homocytotropic antibody in sheep. *Immunology,* 16, 543.

HOGARTH-SCOTT, R.S., JOHANSSON, S.G.O. & BENNICH, H. (1969) Antibodies to *Toxocara* in the sera of visceral larva migrans patients: the significance of raised levels of IgE. *Clin. exp. Immunol.* 5, 619.

HUDSON, R.J., BANDY, P.J. & KITTS, W.D. (1970) Immuno-chemical quantitation of ovine immunoglobulins. *Amer. J. vet. Res.* 31, 1231.

HUDSON, R.J., KITTS, W.D. & BANDY, P.J. (1970a) Monitoring parasite activity and disease in the Rocky Mountain bighorn by electrophoresis of seromucoids. *J. Wildl. Dis.* 6, 104.

HUDSON, R.J., KITTS, W.D. & BANDY, P.J. (1970b) Homocytotropic and blocking antibodies in lungworm

(*Protostrongylus* sp.) infections of the Rocky Mountain bighorn sheep. Paper presented to the Canadian Society of Animal Production, Parksville, British Columbia, 16–18 June, 1970.

ISHIZAKA, K. & ISHIZAKA, T. (1967) Identification of γ E-antibodies as a carrier of reaginic activity. *J. Immunol.* **99**, 1187.

ISHIZAKA, K. & ISHIZAKA, T. (1968) Reversed-type allergic skin reactions by anti-γ E-globulin antibodies in humans and monkeys. *J. Immunol.* **100**, 554.

ISHIZAKA, T., ISHIZAKA, K., ORANGE, R.P. & AUSTEN, K.F. (1970) The capacity of human immunoglobulin E to mediate the release of histamine and slow reacting substance of anaphylaxis (SRS-A) from monkey lung. *J. Immunol.* **104**, 335.

JESKA, E.L. (1969) Mouse peritoneal exudate cell reactions to parasitic worms. I. Cell adhesion reactions. *Immunology*, **16**, 761.

JOHANSSON, S.G.O. & BENNICH, H. (1967) Immunological studies of an atypical (myeloma) immunoglobulin. *Immunology*, **13**, 381.

KAPLAN, A.M. & FREEMAN, M.J. (1969) Occurrence of ovine homocytotropic antibody. *Proc. Soc. exp. Biol. (N.Y.)*, **132**, 514.

LAYTON, L.L. (1965) Passive transfer of human atopic allergies into lemurs, lorises, pottos, and galagos: possible primate-ordinal specificity of acceptance of passive sensitization by human atopic reagin. *J. Allergy*, **36**, 523.

LEVY, D.A. & OSLER, A.G. (1967) Studies on the mechanisms of hypersensitivity phenomena. XVI. *In vitro* assays of reaginic activity in human sera: effect of therapeutic immunization on seasonal titre changes. *J. Immunol.* **99**, 1068.

MCANINCH, J.R. & PATTERSON, R. (1970) Reagin-like antibody formation in rabbits during a heterogeneous antibody response to *Ascaris* antigens. *Immunology*, **18**, 91.

MONGAR, J.L. & WINNE, D. (1966) Further studies on the mechanism of passive sensitization. *J. Physiol. (Lond.)*, **182**, 79.

MORSE, H.C., AUSTEN, K.F. & BLOCH, K.J. (1969) Biologic properties of rat antibodies. *J. Immunol.* **102**, 327.

MORSETH, D.J. & SOULSBY, E.J.L. (1969) Fine structure of leukocytes adhering to the cuticle of *Ascaris suum* larvae. II. Polymorphonuclear leukocytes. *J. Parasitol.* **55**, 1025.

MOTA, I., SADUN, E.H., BRADSHAW, R.M. & GORE, R.W. (1969) The immunologic response of mice infected with *Trichinella spiralis*. Biological and physico-chemical distinction of two homocytotropic antibodies. *Immunology*, **16**, 71.

OGILVIE, B.M. (1944) Reagin-like antibodies in animals immune to helminth parasites. *Nature (Lond.)*, **204**, 91.

OLIVEIRA, B., OSLER, A.G., SIRAGANIAN, R.P. & SANDBERG, A.L. (1970) The biologic activities of guinea pig antibodies. I. Separation of γ_1 and γ_2 immunoglobulins and their participation in allergic reactions of the immediate type. *J. Immunol.* **104**, 320.

PAN, I.C., KAPLAN, A.M., MORTER, R.L. & FREEMAN, M.J. (1968) Spectrum of ovine immunoglobulins. *Proc. Soc. exp. Biol. (N.Y.)*, **129**, 867.

PARISH, W.E. (1970) Absorption of reagin by human tissues, *in vitro*. *Int. Arch. Allergy*, **37**, 184.

PATTERSON, R. (1969) Laboratory models of reaginic allergy. *Progr. Allergy*, **13**, 332.

REVOLTELLA, R. & OVARY, Z. (1969) Reaginic antibody production in different mouse strains. *Immunology*, **17**, 45.

SCHEIDEGGER, J.J. (1955) Une micro-méthode de l'immunoélectrophorèse. *Int. Arch. Allergy*, **7**, 103.

SOULSBY, E.J.L. (1967) Lymphocyte, macrophage, and other cell reactions parasites. Proc. Spec. Sess. 6th Meet. PAHO Advisory Comm. Med. Res. Pan American Health Organization, Washington, D.C. Scientific Publication No. 150, 66.

STREJAN, G. & CAMPBELL, D.H. (1968) Hypersensitivity to *Ascaris* antigens. IV. Production of homocytotropic antibodies in the rat. *J. Immunol.* **101**, 628.

WERNER, M. (1969) Serum protein changes during the acute phase reaction. *Clin. chim. Acta*, **25**, 299.

206

Nutrition and Parasitism Among Rural Pre-School Children in South Carolina

J. P. CARTER, M.D., Dr. P.H.

R. VANDERZWAAG, Ph.D.

W. J. DARBY, M.D., Ph.D.

E. J. LEASE, Ph.D.

F. H. LAUTER, Ph.D.

B. W. DUDLEY, M.A.

E. G. HIGH, Ph.D.

D. J. WRIGHT, M.D.

T. MURPHREE

IN 1963 Jeffrey et al. reported the prevalance of intestinal helminth infection in Beaufort and Jasper counties along the coast in South Carolina.[1] Two hundred and twelve persons of all age groups were examined. Sixty-four per cent were infected with *Ascaris lumbricoides* and 37 per cent with *Trichuris trichiura*. For the Ascaris infections, the average egg-count was 45,000 per gram of feces. This area of South Carolina is a rural coastal area, south of Charleston near the Georgia border. Many of its inhabitants live in dilapidated shacks with no, or inadequate, sanitary facilities. Two of the poorer homes are depicted in Figure 1. In some cases, families did not have outdoor privies.

In January 1968, Bonner of the U.S. Navy, Parris Island, repeated the survey for intestinal helminths.[2] This second survey was done on behalf of Penn Community Services, Inc., Frogmore, South Carolina, to help them make recommendations to the Beaufort County Health Department. Eighty pre-school children from the two communities of Big Estate and Pritchardville, located at opposite ends of Beaufort County were selected. The results of this study were essentially the same as the earlier one. Fifty-four and one-half per cent were infected with either Ascaris or Trichuris or both.

The present study was undertaken to determine the nutritional status of pre-school children from low income families living in the Beaufort-Jasper areas, and to determine the interrelationships between nutrition and parasitism. Qualitative data on dietary intakes and intestinal parasites and

Fig. 1. Sub-standard housing and living conditions.

hemoglobin and vitamin C levels on these children have been published in a preliminary report by Lease, et al.[3] The initial vitamin A levels in these children have been reported by High[4]. This report represents the completed study.

MATERIALS AND METHODS

The subjects were 178 pre-school Negro children between the ages of two and eight years and 46 white pre-school children who were residents of the Bluffton-Hilton Head Island area of South Carolina. The numbers of children in each age category were similar in both the Negro and white groups. The names of these children were taken from the files of the local well-baby clinics. Mothers of these children, therefore, at one

time or another had come to the health center seeking medical supervision for their children. Only children living in the Bluffton and Hilton Head areas were included in the study.

It is generally agreed that because of the private resort industry on Hilton Head Island, the standard of living and socio-economic conditions, including housing, are better here than in some of the other regions of Beaufort and Jasper Counties. The sample, therefore, is not representative of all black and white pre-school children from low-income families in Beaufort and Jasper Counties. If anything, the taking of the sample from the files of the health centers tended to select those children of better nutritional and health status because their parents were interested in their health and were at least aware of the existing free health services.

The children were examined during the summer of 1968 and in December 1968. They were brought to the public health centers at Hilton Head and Bluffton and were also examined in their homes. A clinical nutritional examination was done similar to the one outlined in the Interdepartmental Committee on Nutrition for National Defense, Manual for Nutrition Surveys (ICNND).[5] Anthropometric measurements including height, weight, head circumference, chest circumference, and subcutaneous fat, as determined by skin-fold thickness in the mid-triceps and subscapular areas, were also recorded. Only the Negro children had physical and biochemical examinations. The white children had their dietary histories recorded and stool examinations performed.

Fecal samples were examined for ova and parasites, using the direct smear technic and a modified zinc sulfate flotation method.[6] Egg-counts using the method of Stoll and Hausheer[7] were done on a subsample of 30 children selected at random from those positive on direct smear to determine the worm burdens for Ascaris and Trichuris. Soil samples from the door yards of about a dozen homes of infected children were examined for parasite eggs by modifying the sedimentation test of Headlee.[8]

Using a 24-hour recall technic[9], dietary information was obtained on each child by interviewing mothers and guardians. The daily intake of specific nutrients was then calculated by referring to the United States Department of Agriculture dome and Garden Bulletin No. 72. "Nutritive Value of Foods." The adequacy of the diets was determined by comparison with the Recommended Dietary Allowances of the National Research Council published in 1964.[10]

Blood samples were obtained on 48 children during the summer and on 101 children in December of 1968. The summer and winter values were done on the same children. Hemoglobin values were determined by the cyanmethemoglobin method.[11] Serum Vitamin C levels were determined by using the dinitrophenylhydrazine method.[5] Total protein was determined by the standard biuret method.[12] Electrophoretic patterns were determined using a cellulose acetate membrane and a Beckman microzone electrophoresis apparatus.[13]

Vitamin A and serum carotene were determined by the trifluoroacetic acid method.[6] Serum iron and total iron binding capacity were determined by the Ramsey method.[14] Serum folate was determined by microbiological assay.[15] Alkaline phosphatase was determined by the King-Armstrong method.[16] Transketolase and thiamine pyrophosphate effect were determined as outlined by Brin.[17]

In addition a detailed questionnaire was designed and used to determine the family composition, socio-economic status, parental educational level, housing, sanitary facilities and overall living conditions of the families examined. Information was also obtained on utilization of existing health services, food habits and the food purchasing patterns of the families in question. Data obtained from this questionnaire were then summarized and correlated with nutritional status and the presence or absence of intestinal parasites. The questionnaire was only administered to blacks and not to whites.

RESULTS

An analysis of the heights and weights of the children examined are presented in Figures 2 and 3. The data are for 178 Negro children and show that the observed distribution curves for both height and weight for boys and girls are significantly different from the expected ones, which are based on the Iowa-Boston norms (P<0.005).

The skin-fold thickness measurements are also significantly different from the standards for subcutaneous fat in British children compiled by

Tanner.[18] These data are presented in Figures 4 and 5.

There were 19 children out of 157 whose head circumference was bigger than the chest circumference. It had been postulated by the late R.F.A. Dean that if the head circumference is bigger than the chest circumference in a child after six months of age, the child is malnourished 95 percent of the time.[19]

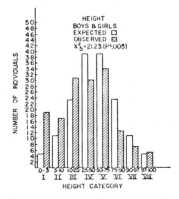

Fig. 2. Observed and expected numbers of children in various height percentile categories. (Based on the Boston norms.)

The dietary data of the Negro and white children are presented in Table 1. The most limiting nutrients as far as the Negro children are concerned are calories, iron and vitamin A. The diets of the white children are better in respect

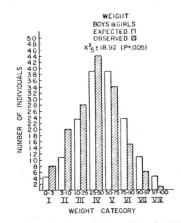

Fig. 3. Observed and expected numbers of children in various weight percentile categories (Based on the Boston norms.)

to every single nutrient except vitamin A. They are limiting in iron as well. The high intake of protein in both groups is due to the consumption of shrimp and fresh fish in this coastal area. The higher intake of vitamin A in the Negro group is probably due to the consumption of greens. The diets of the Negro children differ only from those of pre-school children in developing countries in the higher intake of protein.

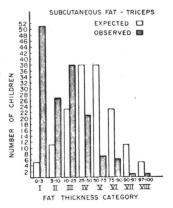

Fig. 4. Observed and expected numbers of children in triceps skinfold thickness category. (Based on the English norms.)

The clinical findings are tabulated in Table 2. The most outstanding clinical signs of malnutrition and parasitism in the Negro children were dyspigmentation of the hair in 25 per cent, easily pluckable hair in 35 per cent, dry or scaling skin

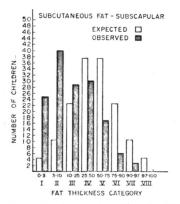

Fig. 5. Observed and expected numbers of children in the sub-scapular skin-fold thickness categories. (Based on the English norms.)

in 13 per cent, pallor of the fingernails in 18 per cent, cheilosis of the lower lip in 28 per cent, periodontal disease in 30 per cent, dental caries in 50 per cent, filiform atrophy of the tongue in 16 per cent, distended abdomen in 41 per cent, bowing of the legs in 2 per cent, parietal protruberance in 15 per cent, and bossing of the skull in 4 per cent. In addition, 22 per cent of the children were found to have an umbilical hernia. The high incidence of umbilical hernia in Negro children has been reported elsewhere.[20] These findings are suggestive of iron deficiency anemia, dental caries, periodontal disease, rickets, and malabsorption. The malabsorption, suggested by a distended abdomen, is probably secondary to infection with Ascaris.

TABLE 1.—AVERAGE DAILY DIETARY INTAKES

Nutrient	Group	
	Negro	White
Calories	888	1369
Protein (gms)	46	66
Calcium (mg)	380	510
Iron (mg)	5.1	7.2
Vitamin A (I.U.)	2243	1911
Vitamin B₁ (mg)	.46	.60
Vitamin B₂ (mg)	.79	1.40
Niacin (mg)	6.96	10.69
Vitamin C (mg)	32	47

Blood specimens were drawn on the Negro children. The mean, range, and standard deviations for the hemoglobin and the other biochemical parameters are given in Table 3. In Table 4, the number and per cent of children falling outside of the normal range are also given. The data are broken down by season, i.e. summer of 1968, December of 1968, and both seasons combined.

The results show that 26 per cent of the children were anemic with hemoglobin levels below 10 gms. per 100 ml. The discrepancy in the hemoglobin values between summer and winter is due to the fact that difficulty was encountered by the medical students in obtaining blood from the children under two years of age. Thirty per cent of the children had low levels of vitamin A in the serum, that is to say, the levels were below 20 μg per cent. This is the level below which night blindness occurs (Sauberlich, unpublished observations). Fourteen per cent of the

TABLE 2.—CLINICAL FINDINGS IN NEGRO CHILDREN

	Per cent
Hair	
Dry	16.56
Dyspigmented	24.53*
Pluckable	34.91*
Abnormal texture or loss of curl	1.91
Fungus	0.94*
Skin	
Dry or scaling (xerosis)	13.38
Hyperpigmentation of face or hands	1.27
Thick pressure points	1.27
Fungus	5.10
Nails	
Pallor	17.92*
Clubbed	5.66*
Eyes	
Conjunctival injection	9.55
Lips	
Cheilosis	28.30*
Gums	
Periodontal disease	30.18*
Teeth	
Missing teeth	0.94*
Carious teeth	50.00*
Fluorosis	0.64
Tongue	
Papillary atrophy (fungiform)	2.55
Smooth periphery (filiform atrophy)	16.3*
Fissured	0.64
Glands	
Palpable thyroid	84.91*
Visible, head extended	36.79
Abdomen	
Prominent	40.57*
Hepatomegaly	10.19
Splenomegaly	—
Diastasis Recti	5.10
Umbilical Hernia	21.70*
Skeletal	
Chest deformity	0.94*
Beading of ribs	0.64
Bowed legs	1.88*
Epiphyseal enlargement	1.97
Bossing of skull	3.77*
Flat occiput	—
Parietal protuberance	14.65

* Percentage based on sample size of 106 (all other percentages are based on sample size of 157).

children had low levels of folic acid; they had levels less than 4 mμg per cc.

The mean ascorbic acid level was actually higher in the winter than in the summer. However, there were two children out of 99 with dangerously low levels (below 0.19 mg per cent) in the winter. One child had a level as low as 0.09 mg per cent. The serum carotene level was definitely higher in the summer than in the winter. There was no significant difference between the vitamin A level in the summer and winter.

There were deficiencies of vitamin B_1 or thiamine in 8 per cent as determined by transketolase levels and the effect of stimulation with thiamine pyrophosphate, i.e. both were abnormal in 8 per cent. The serum enzyme alkaline phosphatase was determined and was above 19 units in 26 per cent. Interestingly enough, the alkaline phosphatase levels were extremely low (less than 4 K.A. units) in 71 per cent. The reasons for these extremely low values for alkaline phosphatase on the one hand, and the high values on the other are obscure. It is possible, that the high values represent a compensatory mechanism to increase bone calcification because of inadequate amounts of vitamin D. The low values could possibly be explained on the basis of a primary deficiency of zinc.[21] There was no correlation between the alkaline phosphatase values and height.

The serum iron level was below 40 μg per cent in 51 per cent of the children examined. The per cent saturation was 10 per cent and less in these cases. The anemia which was found in one quarter of the children, therefore, is probably the result of the combined deficiencies of iron and folic acid. Iron deficiency, however, is probably the greatest single contributor, and is undoubtedly related to both inadequate intake and chronic blood loss, the latter being secondary to infection with large numbers of *Trichuris trichiura*.

Total protein and serum albumin were both within normal limits. The mean gamma globulin was slightly elevated and 17.4 per cent of the total proteins. Otherwise, protein electrophoresis was normal.

The number and per cent of Negro and white children with intestinal parasites are given in Tables 5 and 6, respectively. Four per cent of the white children and 73 per cent of the Negro children were infected with either Ascaris or Trichuris, or both.

TABLE 3.—BIOCHEMICAL FINDINGS IN NEGRO CHILDREN

	Mean	Low	High	Standard Deviation
Hemoglobin (gm/100ml)				
Summer	11.70	7.5	13.9	1.25
Winter	10.46	5.0	12.8	1.25
Ascorbic acid (mg/100ml)				
Summer	1.02	0.40	1.84	0.32
Winter	1.22	0.09	3.77	0.75
Carotene (mg/100ml)				
Summer	121.65	52.4	217	32.80
Winter	85.51	1.24	213.8	41.36
Combined	97.41	1.24	217.0	42.85
Vitamin A (μg/100ml)				
Summer	25.32	1.6	46.8	10.45
Winter	27.80	7.3	69.4	14.9
Combined	26.93	1.6	69.4	13.72
Serum Fe (μg/100ml)—Winter	43.92	1	162	34.26
TIBC (μg/100ml)—Winter	457.21	143	864	152.0
Folate (mμg/cc)—Winter	6.67	2.8	17.3	2.92
Transketolase (units)—Winter	726.80	481	1148	129.14
Tpp Effect (%)—Winter	6.22	0.0	26.0	5.77
Alk. Phos. (K.A. units)—Winter	8.31	0.0	34.0	11.66
Serum albumin (gm/100ml)—Summer	4.58	3.5	6.9	0.62
γ-Globulin (%)—Summer	17.36	9.7	26.9	4.62
Total serum pteteins (gm/100ml)				
Combined	7.56	5.6	10.0	0.71

The distribution of egg-counts for a sub-sample of 30 of these infected children were selected randomly. What these mean in terms of estimates of worm burdens and losses of blood, calories, and protein were determined. The approximate number of worms, based on equal numbers of males and females, was determined by calculations based on figures compiled by the Communicable Disease Center, Atlanta, Georgia, from the data of Stoll and Beaver[22, 23]. The amounts of calories consumed by the Ascaris worms were estimated from the data of von Brand.[24] Protein loss due to defective digestion and absorption secondary to Ascaris infection was based on the data of Venkatachalam and Patwardhan.[25] In addition, a second set of estimates as to the number of Trichuris worms and the blood loss resulting therefrom were determined. The figures determined are based on the more recent estimates of Layrisse et al.[26-28] These data state that 800 parasites are equivalent to 15,000 eggs/gm. feces.

The number of Trichuris worms are based on the older data of Stoll and Beaver compiled by the C.D.C. which states that 20,000-50,000 Trichuris eggs/gm. is equal to 160-400 worms. In any case, these techniques do not give exact counts of the number of worms present. The number of eggs per gram in a stool specimen will depend on diet, where the worms live in the intestinal tract, stool consistency, and the variation in the daily output of feces.

A number of studies have shown that Ascaris infection may contribute to Vitamin A deficiency. Children suffering from night blindness have been found to show rapid improvement in their eye symptoms within a few days of elimination of the parasite.[29] In Table 7, therefore, the relationship between infection with Ascaris lumbricoides and serum Vitamin A levels in 50 subjects is shown. There is no significant correlation.

Soil samples from door yards of about a dozen homes of infected children were examined, and all were found to be contaminated with parasitic eggs.

The socio-economic questionnaire was designed to obtain data in the following areas: 1) family composition, 2) educational level of the parents or guardians, 3) health and dental care, 4) nutrition, 5) condition of the house and related questions, 6) household appliances, 7) water, sanitation, and hygiene and 8) income and savings.

The questionnaire was administered to 105 black families. The answers to specific questions

TABLE 4.—ABNORMAL VALUES OF BIOCHEMICAL FINDINGS IN NEGRO CHILDREN

	Summer No.	%	Winter No.	%	Combined No.	%
Hemoglobin <10 gm/100ml	2	4.55	26	25.74		
Ascorbic Acid <0.3mg/100ml	0.	0.0	2	2.02		
Vitamin A <20µg/ml	12	24.0	24	30.0	35	30.4
Serum Fe <40mµg/100ml			45	51.14		
Folate <4mµg/ml			13	13.98		
Transketolase <750 units			55	56.7		
Tpp Effect >15%			8	8.25		
Alk. Phos.						
>19 K.A. units			19	26.39		
< 4 K.A. units			51	70.83		
Total serum proteins <6 0 gm/100 ml					1	1.20

are given below. More importantly, however, the data obtained summarizes in a general way the living conditions of the low-income Negro families in the Beaufort and Jasper Counties area.

In answer to the question, "Who in the house is the child's mother or functions as the mother?", 87 per cent of the respondents replied the child's natural mother. The natural mother was at home one-half of the time or more in 88 per cent of the cases. Twenty-nine per cent of these mothers completed less than the seventh grade in school; about one-third got as far as the seventh to the ninth grade; only 21 per cent of the mothers completed high school.

In 63 per cent of the cases, the father in the house was the child's natural father. The natural father was at home one-half of the time or more in 53 per cent of the families questioned. In 24 per cent of the cases, it was not known how far the father got in school. In 24 per cent of the cases, the father had completed less than the seventh grade. Only 11 per cent of the fathers had completed high school. In 57 per cent of the cases, the natural father was the head of the household.

In the area of utilization of health services, 58 per cent of the respondents replied that they only take their children to the doctor when they are sick. Ninety-seven per cent of the families had never taken their children to see a dentist. In one-third of the households, neither the mother nor the father had any health insurance coverage. In 45 per cent of the families, hospitalization insurance was available. This hospitalization coverage did not include the children. In 72 per cent of the cases, the children had no health insurance coverage whatsoever.

The questions on nutrition revealed that one-fifth of the households, when asked if the child had consumed foods from the four basic food groups, i.e. bread/cereal, dairy products, vegetables/fruit, and meats/fish, during the preceding day, replied that the child had less than three groups represented in the diet throughout the day. One-fifth of the respondents replied that the child had only two meals with three groups represented throughout the preceding day. In 24 per cent of the cases, the child had three meals with all four food groups present somewhere during the preceding day. Only in 24 per cent of the cases, did the child get a hot lunch at school. In 87 per cent of the cases, the families were not receiving food stamps.

The questions relating to the condition of the house revealed that in 19 families out of 105, there were 11 or more people living in the house; in an additional 44 per cent of the families, seven to 10 people were living in the house; in only two families were there less than three people living in the house. The average number of rooms in the homes was between four and five. The protection quality of the house as judged by the walls and insulation was poor in 40 per cent of the cases. The outside of the house had no paint or varnish in 26 per cent of the cases. The condition of the roof of the house was considered poor and worn with leaks in about 23 per cent of the cases. The condition of the floor was considered poor with cracks and holes in 30 per cent of the cases. In 12 per cent of the cases, the windows were of broken glass, uncovered, and there were no screens or shutters. The outside doors of the house were all there but in poor condition in 30 per cent of the cases

In most cases, i.e., 50 per cent, the houses were heated by a wood stove other than the kitchen stove or a space heater; only 40 per cent of the homes had gas, oil, coal furnace, or an electric furnace. Nearly all of the families had electric lights.

In one-fifth of the cases, the father and/or mother owned the house and there was no mortgage; in one-third of the cases they owned the house, but it was mortgaged; in one-third of the

TABLE 5.—INTESTINAL PARASITES IN PRE-SCHOOL NEGRO CHILDREN BLUFFTON-HILTON HEAD ISLAND, S.C. AREA

	Number	Per cent
Infected with Ascaris only	28	21.4
Infected with Trichuris only	33	25.2
Infected with both Ascaris and Trichuris	35	26.7
Not infected with Ascaris or Trichuris	35	26.7
Total Number of Stools Examined	131	100.0
Total Number of Children Infected with Parasites	96	73.3

cases the house was being rented. Thirty-nine per cent of the respondents said that they were also renting the land; 10 per cent said they owned the land and were making payments; 20 per cent said they owned the land and had paid off the necessary payments. Seventy per cent of the families had less than five acres; 29 per cent had from one to five acres; and 41 per cent had less than one acre. An automobile was owned and usable in 46 per cent of the cases; in another 17 per

TABLE 6.—INTESTINAL PARASITES IN WHITE PRE-SCHOOL CHILDREN BEAUFORT COUNTY, DEC. 1968

	Number	Per cent of Total Examined
Infected with Ascaris only	1	2.2
Infected with Trichuris only	1	2.2
Infected with both Ascaris and Trichuris	0	0
Not infected with Ascaris or Trichuris	44	95.6
Total number of stools examined	46	100.0
Total number of children infected with Parasites	2	4.4

cent of the cases, the automobile was not owned but was available; in one-quarter of the cases an automobile was not owned and none was available.

In the case of household appliances, nearly every family had a refrigerator. Only one-third of the families had a freezer. In 97 per cent of the cases the cooking stove was gas or electric. One-half of the families did not own a washing machine; one-third owned a non-automatic, non-electric washing machine. Fifty-eight per cent of the families had both a radio and a television set.

TABLE 7.—RELATIONSHIP BETWEEN SERUM VITAMIN A LEVEL AND ASCARIS INFECTION

	Vitamin A < 20 μg.	Vitamin A > 20 μg.	Totals
With Ascaris Infection	9	12	21
Without Ascaris Infection	6	23	29
Totals	15	35	50

When questioned about where the child included in our study slept most of the time, one-half of the respondents said that the child shared a mattress, bed, or cot with one or two other children. In 36 per cent of the cases, the child shared a mattress, bed or cot with one adult or more than two other children. In 45 per cent of the cases, three to four people slept in the same room with the child. In only three instances did the child have a room by himself.

In 59 per cent of the homes, the water came from a shallow well. In no cases was city water supplied. In one-third of the cases, the water came from a deep well. In 60 per cent of the cases, there was no running water in the house. Only 30 per cent of the families had a bathtub or shower. One-third of the families had a toilet inside the house; 40 per cent had a toilet or latrine outside of the house; and one-quarter of the families did not have a toilet or latrine inside or outside of the house.

In answer to the question, "How much money did the family make last year?", 41 per cent replied over $3,000; 20 per cent replied between $2,400 and $3,000. The remainder, with the exception of three families, earned less than $2,400 per year. In answer to the question, "Where did most of the money come from?" 55 per cent said the money came from regular, full-time working; in 18 per cent of the cases, the money came from seasonal and/or part-time working. In only 12 cases (11 per cent), did the money come from welfare and old age payments; in 13 cases (12 per cent), the money came from social security and workmen's compensation. In three-quarters of the cases, therefore, 75 per cent or more of the income came from wages. In 13 per cent of the cases, there were 11 or more people sharing the father's and/or mother's income. In 49 per cent of the cases, there were seven to 10 people sharing the family income. In 19 per cent of the cases, there were three to four people sharing the family income.

In answer to the question, "How much of the money goes for food?", 38 per cent of the families replied between 17-34 per cent. One-fifth of the families replied between 34 percent and 50 per cent.

In answer to the question, "About how much of the money goes for rent?", 42 per cent replied between 0 and 17 per cent; 26 per cent of the families replied between 17 per cent and 34 per cent; 16 per cent of the families replied between 34 per cent and 51 per cent. In answer to the question, "Have you been able to put away any money for a rainy day?", 73 per cent of the respondents replied, No.

We examined in greater detail the questionnaires of 10 children considered to be normal, or as close to normal as possible, from an anthropometric, clinical, biochemical, and parasitological standpoint. These 10 children were compared with 10 children who were parasitized and/or malnourished. Only children from different families were included in the comparison. The numbers of children in each category were not sufficiently large enough to permit a completely random selection. In general, the data revealed no differences in the questionnaires of the families of the parasitized and/or malnourished children and the normals. For example, six mothers in the normal group completed high school, but four mothers in the parasitized and/or malnourished group got as far as the 11th grade and an additional two completed high school. Six mothers in the normal group said they took their children to the doctor only when they were sick, four mothers in the parasitized and/or malnourished group said so. Two mothers in the normal group and two in the

parasitized and/or malnourished group had no toilet or latrine; five of the normal and four of the parasitized and/or malnourished had a tiolet inside the house. Four families in the normal group and five in the parasitized and/or malnourished group earned over $3,000 per year. In eight and six families, respectively, the money came from regular, full-time working.

Beaufort and Jasper Counties are two of the poorest counties in South Carolina. Data compiled by Coles and Huge show that "there are 17,536 people in Beaufort County with an income under $3,000 a year; 11,064 see less than $2,000 a year; 1,145 families get less than $1,000 per year; only 898 people receive public assistance; in 1965 the infant mortality rate, among the very highest in America, was 62.4 per 1,000 live births; about four per cent of Beaufort's people have health insurance; about half the county's people have less than eight years of education, and 20.4 per cent have been declared 'functional illiterates'; over a third live in houses called 'substandard'."[30]

The results of this study would seem to indicate that for many of the black families in this area, housing is inadequate and community water supplies and sewerage and waste disposal systems are practically non-existent. The socio-economic data on these families are surprising in that they destroy some of the myths about the Negro family in the South. In the majority of the homes, both the mother and the father were present. Only 10 per cent of the families were on welfare. Three-quarters of them earned their income by full-time, or seasonal employment. The wages they earned for unskilled labor were minimal however. The majority of the heads of households had at least an eighth grade education.

The children of these families show dietary, clinical and biochemical evidence of malnutrition. They also show a high prevalence rate of infection with the intestinal round worms *Ascaris lumbricoides* and *Trichuris trichiura*. Both malnutrition and parasitism act synergistically. A good example of this is the fact that 50 per cent of the children were suffering from systemic iron deficiency. It is known that the average blood loss produced by a single Trichuris worm is 0.005 cc of blood per worm per day.[31] This chronic blood loss, therefore, contributes to the iron deficiency and the anemia.

In heavy infections, it can be as much as four to 26 cc of blood per day. Both systemic iron deficiency and anemia are also likely to develop when the average daily intake of iron is less than $\frac{1}{2}$ of the recommended allowance, as in the case of the Negro children in this study where it was five mg. per day.

Protein losses also occur in Ascaris infection because of poor digestion and absorption. The worm is known to secrete an antitrypsin and antipepsin enzyme.[31] It is doubtful that this parasite consumes enough food to interfere with the nutrition of the host. According to von Brand,[24] "50 ascarids would withdraw only $\frac{1}{2}$ of 1 per cent of a 2,500 calorie-diet, or perhaps 1 per cent of a starvation diet." It can cause damage, however, by interfering with the digestion and absorption of protein, migrating to aberrant sites such as the biliary tract and tracheo-bronchial tree, and balling itself together in sufficient numbers to cause intestinal obstruction.

Our estimates as to the numbers of parasites harbored by infected children in this study indicate that one-quarter of those infected had heavy infections with Trichuris (800 worms or greater), and nearly one-half (42 per cent) of those infected with Ascaris had moderate to heavy infections with Ascaris (10 worms or greater).

The end result of this chronic state of undernutrition and intestitinal parasitism can be seen in the anthropometric measurements recorded on these children. There is evidence of growth retardation and decreased amounts of subcutaneous fat.

Perhaps even more important, however, are the effects of deprivation which result from living and growing up under conditions such as those described in the results of the socio-economic survey. These conditions are the result of a number of environmental factors which have long existed. The lack of a potable water supply and waste disposal systems are degrading in this affluent society.

SUMMARY

One hundred and seventy-eight pre-school Negro children were examined in Beaufort and Jasper Counties in South Carolina. The sample was selected by taking available children on Hilton Head Island and on the nearby mainland. Their names were taken from the files of the local well-baby clinics. Significant numbers of these children were found to have heights and weights below the norms for their chronological ages. Many of them had decreased amounts of subcutaneous fat as measured by skin-fold thickness. There were dietary, clinical, and biochemical evidences of caloric deficiency, iron deficiency, folic acid deficiency, and Vitamin A deficiency.

Seventy-three per cent of the children were found to be infected with either *Ascaris lumbricoides* or *Trichuris trichiura* or both. Egg-counts done on a sub-sample of 30 of the infected children showed that 42 per cent of the children infected with Ascaris had over 10 worms and 27 per cent of the children infected with Trichuris had over 800 worms.

Both poor nutrition and intestinal parasitism, therefore, were acting synergistically to impair the growth and development and health of the subjects examined.

A socio-economic survey showed that poor housing and the lack of a potable water supply and sewerage and waste disposal systems, were additional problems for many of the families living in this area.

ACKNOWLEDGMENTS

We should like to acknowledge the assistance of Dr. Donald E. Gatch and to thank him for making his facilities available. More than any other single person, he inspired this study by calling attention to the fact that malnutrition and parasitism were serious public health problems in the area. We would like to thank the Beaufort County Health Department for space in which to work during the summer phase. We thank the staff of Penn Community Services, Inc., Frogmore, South Carolina, for their help in the preparation of the socio-economic questionnaire. We should like to thank in particular, Mr. J. Golden, Miss P. Weinberg, and Miss C. Schafer for the part they played in designing and administering the questionnaire. Mr. William Deshur, a University of Wisconsin student and two Meharry medical students, Mr. L. E. High and Mr. J. L. Wilson, helped with the physical and stool examinations performed during the summer of 1968. Biochemical determinations were performed by Mrs. J. Hillery and Mrs. E. A. Shute. Mrs. Hillery also helped with the collection of specimens in the field.

The study was made possible by funds supplied by the Field Foundation and funds under Title I of the Higher Education Act of 1965. We sincerely hope that the study will continue to provide a stimulus to work toward improved nutrition and living conditions. The University of South Carolina has inaugurated applied research and educational programs aimed at eradication of intestinal parasites in order to help correct the situations elucidated by this project.

216

LITERATURE CITED

1. JEFFREY, G. M. and K. O. PHIFFER, D. E. GATCH, J. A. HARRISON and J. C. SKINNER. Study of Intestinal Helminth Infections in a Coastal South Carolina Area. Public Health Reports 78 (1), pp. 45-55, Jan. 1963.

2. BONNER, M. Penn Community Services, Inc., Frogmore, South Carolina. Written Report. Jan. 1968.

3. LEASE, E. J. and F. H. LAUTER and B. W. DUDLEY. Intestinal Parasites and Nutritional Status. J. S. Carolina Med. Assn., 65:3, March 1969.

4. HIGH, E. G. Some Aspects of Nutritional Vitamin A levels in Pre-school Children of Beaufort County, South Carolina, Am. J. Clin. Nutr., 22:8, Aug. 1969.

5. Interdepartmental Committee on Nutrition for National Defense. Manual for Nutrition Surveys, 2nd Ed., National Institutes of Health, Bethesda, Md., 1963.

6. FAUST, E. C. and P. R. RUSSELL. Craig and Faust's Clinical Parasitology. Lea and Febiger, Phila. 7th Ed., p. 978. 1964.

7. STOLL, N. R. and W. C. HAUSHEER. Concerning Two Options in Dilution Egg-Counting: Small Drop and Displacement. Amer. J. Hyg., 6:134, 1926.

8. BELDING, D. L. Textbook of Clinical Parasitology. Appleton-Century-Crofts, Inc., New York. 2nd Ed., p. 978. 1952.

9. STEVENS, H. A. and R. E. BLEILER and M A. OHLSON. Dietary Intake of Five Groups of Subjects. J. Amer. Diet. Assn. 42:387, 1963.

10. Food and Nutrition Board: Recommended Dietary Allowances, 7th Revised Edition, 1968. Publ. 1964 Nat. Acad. Sci.-Nat. Research Council, Wash., D.C.

11. HAINLINE, A. et al. Hemoglobin in Standard Methods of Clinical Chemistry. Vol. II. New York, Academic Press, pp. 49-60, 1958.

12. WOLFSON, W. Q. and C. COHN, E. CALVARY and F. ICHIBA. Studies on Serum Proteins, V. A Rapid Procedure for the Estimation of Total Protein, True Albumin, Total Globulin, Alpha Globulin, Beta Globulin and Gamma Globulin in 1.0 ml. of Serum. Am. J. Clin. Path., 18:723-730, 1948.

13. Electrophoresis—Beckman's Instruction Manual RM-IM-3, "A Microzone Method for Serum Proteins Using Ponceau's Fixative-Dye Solution," p. 21-41, Beckman Instruments, Inc., Fullerton, Calif., Aug. 1965.

14. RAMSEY, W. N. M. Clin. Chim. Acta, 2:214, 1957.

15. COOPERMAN, J. M. Microbiological Assay of Serum and Whole-blood Folic Acid Activity. Am. J. Clin. Nutr., 20: No. 9, Sept. 1967.

16. POWELL, M. E. A. and M. J. H. SMITH. J. Clin. Path., 7:1954.

17. BRIN, M. Transketolase and Tpp Effect—Clinical Applications of Transketolase Assay. J. Nutrition, 71:273, 1960.

18. TANNER, J. M. and R. H. WHITEHOUSE: Standard for Subcutaneous Fat in British Children. Brit. Med. J., 1:446, 1962.

19. DEAN, R. F. A. Kwashiorkor. Recent Advances in Pediatrics, Ed. Gairdner, Little, Brown and Co., Boston, 1965.

20. CRUMP. E. P. and J. M. ROBINSON. Umbilical Hernia. J. Ped. 40:777-780, 1952.

21. PRASAD, A. S. and D. OBERLEAS, P. WOLF and J. P. HORWITZ. Studies on Zinc Deficiency. Changes in Trace Elements and Enzyme Activities in Tissues of Zinc-deficient Rats. J. Clin. Invest., 46:549-57, April, 1967.

22. STOLL, N. R. An Effective Method of Counting Hookworm Eggs in Feces, Am. J. Hgy., 3:59-70, 1923.

23. BEAVER, P. C. The Standardization of Fecal Smears for Estimating Egg Production and Worm Burden. J. Parasitol., 36:451-456, 1950.

24. VON BRAND, T. Proc. 4th Intern. Congr. Trop. Med. Malaria, Washington, D.C., Vol. 2: 984-991, U.S. Gov't Printing Office, 1948.

25. VENKATACHALAM, P. S. and V. N. PATWARDHAN. The Role of Ascaris lumbricoides in the Nutrition of the Host. Effect of Ascaris on Digestion of Protein. Trans. Royal Soc. Trop. Med. Hyg., 47:169-175, 1953.

26. ROCHE, M. and M. LAYRISSE. The Nature and Causes of "Hookworm Anemia." Amer. J. Trop. Med. Hyg., 15:1032, 1966.

27. MARTINEZ TORRES, C. and A. OJEDA, M. ROCHE and M. LAYRISSE. Hookwork Infection and Intestinal Blood Loss. Trans. Royal Soc. Trop. Med Hyg., 61:373, 1967.

28. LAYRISSE, M. and L. APARCEDO, C. MARTINEZ-TORRES and M. ROCHE. Blood Loss Due to Infection with Trichuris trichiura. Amer. J. Trop. Med. Hyg., 16:613, 1967.

29. Control of Ascariasis, Report of a WHO Expert Committee, World Health Organization, Tech. Rep. Series, 379, 1967.

30. COLES, R. and H. HUGE. "The Way It Is In South Carolina." The New Republic, 159:17-21, 1968. (Data obtained from County and Data Book. a Statistical Abstract Supplement, Bureau of the Census, U. S. Gov't Printing Office.)

31. Collier. H. B. Canad. J. Res., 19:B19, 90-98, 1941.

217

AUTHOR INDEX

KEY-WORD TITLE INDEX

219